Value Assumptions

in Risk Assessment

A Case Study of the Alachlor Controversy

by Conrad G. Brunk
Lawrence Haworth
Brenda Lee

Wilfrid Laurier University Press

Canadian Cataloguing in Publication Data

Brunk, Conrad G. (Conrad Grebel), 1945-
  Value assumptions in risk assessment : a case
study of the alachlor controversy

Includes bibliographical references.
ISBN 0-88920-200-1

1. Alachlor – Environmental aspects – Canada.
2. Herbicides – Environmental aspects – Canada –
Case studies.   3. Risk assessment – Canada – Case
studies.   4. Environmental impact analysis –
Canada – Case studies.   I. Haworth, Lawrence,
1926- . II. Lee, Brenda, 1951- . III. Title.

TD196.A34B7 1991   383.73'2   C91-094453-9

Copyright © 1991
Wilfrid Laurier University Press
Waterloo, Ontario, Canada
N2L 3C5

Cover design by Connolly Art & Design

Printed in Canada

Value Assumptions in Risk Assessment: A Case Study of the Alachlor Controversy has
been produced from a manuscript supplied in electronic form by the authors.

Contents

Acknowledgements

There are several institutions and individuals whose support and advice were invaluable in completion of this investigation. We would like to acknowledge generous financial support from the Institute for Risk Research at the University of Waterloo. Its Director, Professor John Shortreed, was the first to encourage us to undertake a study of the value assumptions in risk assessment. Grants from the Social Sciences and Humanities Research Council of Canada to two of the authors also made possible the commitment of time necessary for completion of the research and the manuscript.

There are a number of persons whose advice and criticism contributed significantly to our work. Among these are Professor William Leiss of Simon Fraser University, Profesor John Robinson of the University of Waterloo, Professor Peter Wells of the University of Toronto, and Dr. Edwin Levy of QLT Phototherapeutics Inc. All of these people kindly consented to read early drafts of the manuscript and provided us with important critical feedback. None of them should be held in any way responsible, however, for errors of fact or interpretation that remain in the book. For those we assume full responsibility.

This book has been published with the help of a grant from the Canadian Federation for the Humanities, using funds provided by the Social Sciences and Humanities Research Council of Canada.

Introduction: Risk Assessment as Regulatory Science

Debates about risk often take this form: On one side are ordinary citizens, fearful that the asbestos in school walls and ceilings or the radiation from overhanging power lines will kill or maim their children. On the other side are officials of both government and industry who assure the citizens that there is no cause for alarm. Their sanguine view is bolstered by risk assessment experts whom they have called upon for an objective, scientific opinion, based on the actual facts of the case. The alarmed citizens thus tend to be viewed as driven by a well-intentioned but scientifically uninformed concern for the health of their children. They, in turn, may call upon their own group of scientific experts, whose conclusions lend support to their concern.

This raises an important question. Are risk debates disputes between those who accept the findings of science and those who do not? Between good and bad science? Or is it possible that opposing assessments of risk, by scientific experts as well as ordinary citizens, reflect and are guided by dominant values held by the assessors? The following analysis of one of these debates supports the latter view. In it we suggest what those dominant values are, how they work within a risk assessment, and some implications of reconceiving risk debates as primarily debates about values.

* * * * *

Increasingly our society relies upon government regulatory agencies to protect its people, its institutions, and its environment from the negative impacts of new technologies. These agencies are saddled with the task of deciding among strongly conflicting viewpoints rep-

1

resented by a wide range of interest groups and "value communities" within the society. In making regulatory decisions the interests and values of some parties are protected while the interests and values of others are significantly curtailed.

Because the actions of government regulatory agencies profoundly affect the liberties and opportunities of individuals and corporations, the norms of our liberal, democratic social order require that these actions not be arbitrary or capricious. They must be based upon a clearly articulated rationale, and that rationale must not be seriously biased in favour of the values of one interest group against the values of others. That is to say, within the liberal, democratic tradition the rationale by which the government intervenes in the lives of its citizens should either reflect a social consensus or be entirely neutral *vis-à-vis* the competing values and interests affected by it.

The idea that this regulatory process might be carried on by the terms of an objective and socially neutral "science" has become a tantalizing prospect. If science is objective, and if it is value-free, it can arbitrate between competing views about social policy options by demonstrating which of these impose the greatest costs or risks upon the society and generate the greatest compensating benefits; and it can do so by appealing to empirically demonstrable data and principles of science universally accepted even in pluralistic society. Even if it cannot arbitrate among the different *ends* sought by different persons, it can settle disputes about the *means* to the achievement of those ends. If it cannot in the final analysis decide which policy option is the better one, it can still establish objectively the factual implications of each choice.

Hence, the emergence of "regulatory" or "mandated"[1] science—the enlistment of scientists and the trappings of scientific method in the aid of government regulation. An increasingly significant aspect of this regulatory function of governments involves the attempt to protect consumers and the general public from unsafe products and activities, which is to say, products and activities that pose unacceptably high levels of risk. This regulatory function is brought to bear in many ways, including the certification and restriction of potentially harmful products for sale or use in the society, the setting of standards in the design of products, as well as the approval of transportation, energy, and other projects with potential impact upon the social and physical environment.

The expanding awareness of safety issues motivating the rise of government regulatory bureaucracy has also motivated the private sector to employ this "mandated science." In part it has been forced to do so by the public regulatory sector so that it can accommodate its activities to the requirements and standards imposed by the latter.

Both government and industry, therefore, have need of reliable risk assessments to undergird their regulatory or management decisions. Consequently, there has developed a strong market for rigorous and reliable risk assessment and risk management methodologies to aid in these decisions. The more closely these methodologies can approach the ideal of "science," the more they are thought to provide that objective and impartial decision procedure which can serve as the neutral arbiter in the liberal, democratic society.

The emerging discipline of risk assessment/risk management is a response to this need for a regulatory science of safety and risk. Like all regulatory or "mandated" science, risk assessment tries to fill that gap between theoretical or laboratory science on the one hand, and the necessity of making reliable and defensible regulatory or management decisions on the other. In the typical course of events this gap is wide and filled with all manner of obstacles. It is a gap created by the fact that the regulator and the manager do not have the luxury, enjoyed by the laboratory scientist, of waiting until the necessary data are in and all the variables sufficiently controlled for, so that a scientifically reliable conclusion can be drawn. But it is also created by the fact that the kind of question the regulator needs the aid of a risk assessor to answer is not purely scientific.

In the typical case the regulator and the manager have to make a decision about whether to go ahead with, or curtail, some potentially risky plan of action before the data concerning that risk are complete. Often it is not just a matter of waiting for the required information to come in, but rather of having no way to obtain the required information regardless of the time one has. This is particularly the case with respect to risk to human beings. Due to the clear moral difficulties involved in experimenting directly with human beings to see if the product or activity is harmful to them, conclusions have to be extrapolated from data drawn from animal studies or from experience with related compounds in humans. These kinds of difficulties, combined with the fact that the regulator or the manager has to carry out the investigation and the decision in a highly charged political atmosphere, are what create the gap between laboratory science and regulatory or mandated science.

So, even were it true, as many believe, that pure science is an objective and value-free enterprise, it would remain the case that regulatory and mandated science is not. In fact the latter is "science" only in an attenuated sense, notwithstanding the fact that it is typically carried out by scientists, including highly qualified ones.[2]

Although risk assessment is often referred to as one of the new "sciences," and even though its practitioners often invoke the authority of science, it is science only in the attenuated sense of regulatory or mandated science. It approximates the ideal of laboratory science only to the extent that it is able to close the gaps in the data and approach the experimental conditions required by laboratory science. Because of the nature and demands of the regulatory and management situation, however, this gap is hard to narrow, and rarely, if ever, completely closed.

In its classical expressions[3] risk analysis is said to have two stages. The first is a risk estimation stage at which a purely *factual* judgement of the level of risk is made, free of social and other value judgements. It is simply a measure of the magnitude of the harm involved in the event a hazard occurs, multiplied by the probability of its occurrence, both thought to be purely empirical and mathematical problems. Only after an objective assessment of the level of risk has been completed is one qualified to move on to the second stage of risk *evaluation*. This second stage, unlike the first, is admittedly normative; it is the stage at which a determination is made about the *acceptability* of the risk levels identified at the first stage. Since the second stage involves appeal to a norm or standard of acceptability, it is not value-free. But whatever norm of acceptability is appealed to, the resulting safety judgement is deemed by the classical view to be inherently flawed unless it is based upon a scientifically reliable estimate of risk at the first stage.

In the classical model of risk assessment[4] as a scientific enterprise, a distinction is drawn between the objectively determinable, or the real "objective" risk, established by careful use of scientific risk assessment methodology, and the perceptions of risk held by different persons, or the so-called "subjective" risk. The judgements that various persons and interest groups within a society make about the acceptability of risks, for example the risks of generating electricity via fossil-fuel-fired plants in contrast to nuclear generation, are more or less "rational," depending upon how closely they approximate the "objective" determination of risk done by scientific risk estimators. There may be good reasons for making policy decisions affecting the public according to these "irrational," subjective perceptions of risk (reasons having to do with democratic rights, for example), but the perceptions are nonetheless "irrational."

This is how risk assessment is supposed to function ideally as regulatory or mandated science. Our aim in this study is to examine the actual implementation of this risk assessment model in a regulatory situation in order to delineate the assumptions and choices that are at work in the gap between the ideal of laboratory science and the reality of risk estimation in the regulatory situation. We have chosen the case of the registration cancellation of the chemical herbicide alachlor by the Canadian Minister of Agriculture, who determined that it posed too high a risk of cancer to its users. It provides an excellent case study because risk estimation was employed not only in the regulatory decision itself, but even more extensively in an appeal of the government's decision to a review board by the chemical's manufacturer, Monsanto Canada, Inc. The Alachlor Review Board was comprised, with the exception of its chairman, entirely of scientists, and it made the explicit claim that it was doing a purely scientific assessment of the government's cancellation decision.

The Review Board *Hearings*[5] and *Report*[6] provide an excellent case study of the conflicting methodological, conceptual and normative assumptions in risk debates. They are a microcosm of the larger social debates about the assessment and management of risk. Among the witnesses testifying on behalf of the various parties were those defending highly quantitative models of risk estimation (the approach urged by Monsanto) as well as those who advocated other "qualitative" approaches (the view of the government and, ultimately, of the Review Board). Other witnesses representing interested "third parties" in the dispute expressed viewpoints reflecting serious reservations about the whole enterprise of "risk estimation" itself as a scientific or professionalized activity.

In this study we look carefully at the risk estimations of the chemical alachlor offered by the various parties to the dispute, which included the government, the manufacturer, the Alachlor Review Board itself, and several public interest groups. In our estimation the alachlor debate is a typical case of risk assessment employed as regulatory or mandated science. It is a case where despite the amount of test data ultimately made available to the assessors, wide gaps remain, producing uncertainties that open the assessment to a wide range of conflicting conclusions.[7] In the following chapters we attempt to lay bare the implicit value assumptions that motivate and rationalize the choices exercised in these gaps of uncertainty in the data. It is our view that these assumptions are part of an identifiable, coherent framework of social and political values underlying the final assessment offered by those parties who believed that alachlor did not pose unacceptable levels of risk. Those who supported the government's view that the risks posed by alachlor were unacceptable were, on the other hand, guided by different, but equally coherent, value frameworks.

In this, as in many debates about risk and safety, many widely discrepant conclusions were reached about the magnitude and acceptability of the risks. Even parties who claimed to be using objective, value-neutral risk estimation methods and management criteria found themselves in wide disagreement, as often happens among risk assessment experts. If the objectivity of the criteria is assumed, these discrepancies can easily be interpreted as evidence of error by one or more parties in the application of the criteria. However, our analysis of the risk estimations shows that these criteria are not objective and value-neutral, and that these discrepancies are not due to errors of application, but rather to differences in the normative and conceptual assumptions necessarily brought to bear on the process.

Conceptual and normative assumptions are brought to bear not only at the risk *assessment* level, but at the risk *management* level as well. While it is commonly recognized that judgements of safety or risk acceptability are normative, the complexity of the normative issues involved is often overlooked. Indeed, part of the agenda of risk assessment and risk management theory, especially in its most quantitative versions, is to reduce the value components of decisions about safety to a minimum, if not to eliminate them altogether.

There are several sources of the motivation to produce a value-neutral solution to risk and safety problems. One is the widespread view that for a decision to be truly scientific it must not be based on implicit value judgements. The other, and perhaps more influential, motivation is the liberal social ideal that governmental regulatory mechanisms in a pluralistic society be neutral *vis-à-vis* the different goods sought by individuals and groups within society. Thus, decisions about risk and safety are to be made using a methodology that does not discriminate among the goods held. The preference for the risk-benefit standard of determining acceptable risk is motivated by this liberal ideal. It attempts to provide a decision procedure for determining acceptable risks in a society that takes into account equally all social preferences, whatever they may be, without discrimination.

Thus, the risk-benefit standard is a natural aspect of that coherent framework of value which motivates the preference for a "scientific" resolution of risk debates. Indeed, one of the important findings of our study is that this preference for the risk-benefit standard of acceptable risk at the risk management level fed back into and very significantly influenced decisions at the prior risk estimation stage. This finding calls into serious question the classical ideal of risk management based upon a prior, objective estimation of the level of risk.

The conclusion we draw from the alachlor case is that the debate over the risk of alachlor is *not* primarily a debate between those who accept the verdict of the scientific risk estimation and those who do not. It is not a conflict between those who understand the "objective" risks of alachlor and those who are guided by an irrational "subjective" perception of its risks. Neither is it primarily a debate within science itself. Rather, it is primarily a *political* debate—a debate among different value frameworks, different ways of thinking about moral values, different conceptions of society, and different attitudes toward technology and towards risk-taking itself.

If this is true, it raises serious questions about how debates about technological risks ought to be resolved, and whether the appeal to regulatory or mandated science is the best way. It is our view that regulatory sciences such as risk assessment, while they make a valuable, and indeed an essential, contribution, ought not to be the sole arbiters of risk debates. To make them the sole arbiters is in effect to allow one value framework within the social conflict represented in the risk debate to settle the issue in its favour. However, in this way, what is essentially a value-laden political decision becomes disguised as politically and morally neutral. In our view, this is a fundamental reason why appeals to risk assessments as regulatory or mandated science rarely settle risk debates in our society.

I

The Alachlor Controversy

On February 5, 1985, the Canadian Minister of Agriculture cancelled the registration of the herbicide alachlor. Registered by Monsanto Canada, Inc. under the *Pest Control Products Act* in 1969, alachlor had been used by Canadian corn and soybean farmers for nearly 16 years. The decision to reverse the original registration was made by Agriculture Canada upon the recommendation of officials with the Health Protection Branch (HPB) of Health and Welfare Canada after new studies were made available to them indicating that alachlor induced cancerous tumours in laboratory rats and mice during long-term feeding trials.

The 1969 registration had been based upon toxicological studies of the chemical done by Industrial Biotest Laboratories (IBT), an independent testing laboratory in the United States. Investigations of the scandal-ridden IBT conducted by Canadian and American officials in the late 1970s led to the invalidation of many of their pesticide studies, including their studies of alachlor. At the request of HPB Monsanto submitted replacement studies for alachlor in 1982. These studies created "grave concern" about the carcinogenic properties of alachlor, because all but the lowest level of exposure tested in the original studies produced significant carcinogenic responses in the rats used. In addition, studies had come to the attention of the United States Environmental Protection Agency and HPB indicating that alarmingly high levels of alachlor and other pesticides were being found in well water tested in certain farming areas that included Southwestern Ontario.

Health and Welfare officials in the Toxicology Evaluation Division concluded from their review of the information available to them

that "Alachlor is one of the most potent carcinogenic pesticides pres-
ently in use and should be removed from the market as soon as possible" (*Report*, p. 34). A letter was then sent to Monsanto indicating
that Health and Welfare considered alachlor to be a "potential
human carcinogen." In view of the fact that "It is the policy of the
Health Protection Branch to eliminate or reduce to a minimum
human exposure to potential carcinogens," HPB officials indicated
that they were likely to recommend the cancellation of alachlor
(*Report*, p. 35). Following this letter, a series of discussions ensued
between Health and Welfare officials and Monsanto concerning the
different interpretations the two parties made of the data submitted
by the latter.

During 1983, these discussions revolved around additional data
submitted by Monsanto, including results from a second series of rat
studies, skin absorption data, and alternative interpretations of the
original studies. The additional data served to reinforce Health and
Welfare Canada's concerns about the safety of alachlor. After discussing these concerns with officials in Agriculture Canada and the
U.S. Environmental Protection Agency, Health and Welfare notified
Agriculture Canada in June, 1984, that it was its recommendation
that alachlor's registration be cancelled.

Under the terms of the PCPA, the Minister of Agriculture can cancel the registration of a product when, "based on current information
available to him, the safety of the control product or its merit or
value for its intended purposes is no longer acceptable to him" (Sec.
20). Neither the Act nor the Regulations promulgated under it provide the Minister with any guidance as to what levels of safety are or
are not "acceptable," or as to the relationship between safety and
merit (the effectiveness of the product in achieving its claimed objectives) or value (the total benefits derived from the use of the product).

The Act also provides that a company has the right to appeal a
cancellation decision by the Minister to a review board appointed by
the Minister. On March 4, 1985, Monsanto launched such an appeal,
and the following November the Minister appointed a panel of five
persons to hear Monsanto's case for the reinstatement of alachlor.
This Alachlor Review Board held 41 days of public hearings and
heard testimony from over fifty witnesses representing Monsanto,
various government regulatory agencies, farmers' organizations,
environmentalist groups, and private individuals, the written transcripts of which run to nearly 5500 pages. On November 13, 1987,
the Review Board submitted its final report to the Minister of Agriculture, recommending that the registration of alachlor be reinstated.
The Minister, however, did not implement the recommendation, and

at the time of this writing alachlor remains unregistered for sale in Canada.

Summary of the Government's Decision to Cancel Alachlor

In 1982 Monsanto submitted to Agriculture Canada and Health and Welfare Canada studies on alachlor in replacement of the invalid IBT studies that supported alachlor's original registration. These were long-term feeding studies of rats and mice in which the animals were given various doses of alachlor in their food over a period of 25 months. The mouse study showed no significant increase of tumours in the males, but there were statistically significant increases of lung tumours in the females at the highest dose levels. In the rat studies, however, the two highest dose rates produced nasal turbinate tumours and cancers at significant levels, although the lowest dose rate did not.

During 1983, Monsanto submitted additional data, including results from a second series of rat studies, skin absorption data, and alternative interpretations of the original studies. The second series of rat feeding studies used different dosage rates and different terms of exposure to alachlor in the diet. In one study the animals were given the high dose for only 6 months, then fed a regular alachlor-free diet for the remainder of the 25 months. The levels of nasal tumours and cancers found in the animals in this study did not differ significantly from those in which the alachlor had been fed for the full 25 months. This study confirmed the view of HPB officials that there was no reason to assume that the toxic effects of alachlor were a consequence only of long-term chronic exposure to the chemical, but could be induced in humans from the shorter-term exposures to be expected from its use. They consequently rejected Monsanto's argument that the exposures to humans should be "amortized" (the averaging of short-term exposure levels over a total lifetime[8]), thus reducing the estimates of exposure to well below the levels at which toxic effects were observed in the animals.

Several other supplemental studies submitted at this time by Monsanto investigated doses of alachlor at much lower levels. These studies found a statistically significant incidence of nonmalignant nasal turbinate tumours at the 15 mg/kg/day level, and one such tumour at the 2.5 mg/kg/day level. In addition, the studies showed the occurrence of malignant stomach tumours at high dose levels and even one stomach malignancy at the low (2.5 mg/kg/day) level. Although the occurrence of the nasal turbinate and stomach tumours at the 2.5 dose level were not statistically significant, HPB considered them to be *biologically* significant, because of the rare type of the tumours and the fact that their occurrence fitted the pattern established by the higher dose rates.

The appearance of tumours at the low dose rate was especially significant to HPB officials because this rate was within the range of expected dose that HPB had calculated could be expected for the applicators of alachlor. Thus, from their point of view, there was no "margin of safety" at all between potential dose rates experienced by human users of the chemical and those dosages that produced tumours in rats. HPB's calculation of expected exposure rates was based upon the controversial assumptions that the levels of exposure experienced by applicators should not be amortized over a total lifetime, that protective clothing would not be worn or would not be effective in some cases, and that alachlor would be absorbed by the skin at near 100% rates. HPB believed that these were "reasonable" or even "best-case" assumptions, though both Monsanto and the Review Board viewed them as unreasonable worst-case assumptions. The additional data submitted by Monsanto, therefore, served only to reinforce Health and Welfare Canada's concerns about the safety of alachlor.

A controversial issue in the decision to cancel was whether the government considered the risks of alachlor to be unacceptable by the terms of some "absolute" standard of safety independent of other relative risks or benefits, or unacceptable only relative to the risks of alternative weed-control methods. In this case the major alternative control product in use by farmers, and alachlor's major competitor, was the chemically closely related product metolachlor. It was clear that alachlor's cancellation would lead to increased use of metolachlor as a replacement herbicide, so that the natural question, raised by Monsanto as well as government officials, was whether it posed any less risk than alachlor.

HPB officials were concerned about the safety of the alternative metolachlor, and asked that rat studies on its toxicity be reviewed by toxicologists within the Branch. During 1983 and 1984 this review was carried out and HPB informed Agriculture Canada that "repeat chronic studies on metolachlor [have] not demonstrated that this product has carcinogenic activity under the test conditions of these studies even though it is structurally similar to alachlor" (*Report*, p. 39).

HPB officials maintained throughout the Hearings that the question of a safer alternative in metolachlor was not a determinative factor in their recommendation for cancellation, because they considered the risk of cancer posed by alachlor to be sufficient by itself. The Review Board adopted Monsanto's view that the available evidence did not support the government's conclusion that metolachlor was a safer product.

Late in the discussions between the government and Monsanto, the government received new information from the U.S. Environmental Protection Agency concerning the detection of alachlor in water supplies in the U.S. Corn Belt states. This information led Agriculture Canada to implement a small-scale water sampling study in Ontario, which revealed a similar water contamination problem there. These findings were forwarded to HPB and clearly reinforced the concerns there about the risks of alachlor. At the subsequent Hearings before the Review Board, additional well-water studies submitted by the Ontario Ministry of the Environment confirmed HPB's fears that third parties might be exposed to significant levels of alachlor in their drinking water. These studies found alachlor present in 53 out of 351 wells tested in high-use areas of Southwestern Ontario. In some of these wells the levels of alachlor exceeded the Interim Maximum Action Concentration Levels set (later) by Health and Welfare Canada.

The government explicitly rejected the "risk benefit" approach to the question whether the risks posed by alachlor were acceptable. It argued in the Hearings that, in view of the fact that the PCPA authorized the Minister to cancel the registration of a product on the basis of "safety, merit, *or* value," cancellation could be based upon the issue of "safety" alone. This meant, in the government's view, that it was *not* required to balance the benefits (value) of alachlor use against its risks in order to determine the acceptability of the latter.[9] Although the government rejected any *formal* risk-benefit calculus, certain government officials involved in the decision to cancel admitted that benefits were considered and *informally* weighed against the risks (*Hearings*, p. 3451).

The Government rejected formal risk-benefit approaches to the safety issue for several reasons articulated in the Hearings. Wayne Ormrod of Agriculture Canada argued that the agency hesitates to use Risk-Benefit Analysis because it tends to cover up the uncertainties or slant them in one way or another (*Hearings*, p. 3473). He did not elaborate on how this happens. Further, Ormrod pointed out that Risk-Benefit Analysis leaves out important "society values," which ought to remain the "ultimate judgement." By this he apparently meant that the risks "society" might be willing to accept could differ significantly from what a simple weighing of risks and benefits might justify. Ormrod also indicated that the critical issue in the Ministry's judgement against alachlor was not simply the *magnitude* of the risks it imposed but rather the *type* of risk—namely that it was a *carcinogenic* risk (*Hearings*, p. 3509).

Health and Welfare Canada stated in its letter to Monsanto of November 12, 1982, that it was HPB's policy to "eliminate or reduce to a minimum human exposure to potential carcinogens." This is the closest the government came to articulating a standard of acceptable risk throughout the entire process. (It did not repeat or defend this standard in later communications to Monsanto or before the Review Board.) The government's conclusion that alachlor presented an unacceptable risk was, however, entirely consistent with this standard. Given its assumption that there was an alternative product, metolachlor, which could perform as effectively in weed control but without the risk of cancer, the government felt that it could "eliminate" the risk. Even had it concluded that metolachlor, though potentially carcinogenic to humans, was *less* carcinogenic than alachlor, the government could still have justified its decision to cancel by this standard, on the grounds that it was *reducing the risk to a minimum.*

Summary of Monsanto's Case for the Acceptability of Alachlor's Risks

The central questions explicitly raised by Monsanto in its appeal of the cancellation decision were (1) whether the recommendations of HPB and the decision of the Minister were based upon a "reasonable" estimation of the risks posed by alachlor, and (2) whether the government had clearly articulated an appropriate standard of "acceptable risk" and applied it fairly to alachlor and its competitors in the pesticide market.

Monsanto took the position that the long-term rat feeding studies it had submitted to the government did not provide any basis for the latter's conclusion that alachlor posed a risk of cancer to human applicators of the chemical. The company admitted that the studies established the carcinogenicity of alachlor at relatively high doses in rats, but held (a) that there was no reason to extrapolate from the carcinogenic response in the rats to the assumption of such a response in humans, and (b) that even on the assumption that such extrapolation were reasonable, the rat studies suggest that alachlor is not carcinogenic at the levels of exposure to be expected among human applicators and others. The company believed that the fact that there were no statistically significant occurrences of cancers in rats at the low-dose rates indicated a "threshold" level below which alachlor could be assumed to be non-carcinogenic.

The company held that the studies of the probable levels of applicator exposure to alachlor they had submitted showed that these exposures were far lower than those estimated by the government and thus well below the threshold of carcinogenic and oncogenic effect. The company disagreed with the government's assumption of a 100% skin-absorption rate in humans, submitting data from studies with monkeys that showed these animals to absorb on average less than 10% of the alachlor applied to their skin. Further, biomonitoring studies submitted by Monsanto supported their view that far less than full absorption should be assumed. These studies involved analyzing the urine of alachlor applicators to compare the calculated absorbed dose with that obtained from "patch tests" (measuring the amount of alachlor found on patches attached to the skin of applicators). Monsanto argued that these biomonitoring studies (also submitted *after* the Minister cancelled alachlor) supported an estimation of applicator exposure to alachlor 240 times lower than that reached by the government, which had relied solely upon the patch studies.

Monsanto also argued that these exposure levels should be amortized over the lifetime of applicators, since in actual practice they would be exposed to the chemical only five or six days per year. If the exposure levels as calculated by the biomonitoring studies are amortized in this way, Monsanto concluded that the actual expected applicator exposure would be 1,000 to 10,000 times lower than the lowest dose at which tumours were induced in rats.

Monsanto's estimates of applicator exposure were based upon what one of its witnesses called a "typical-case" scenario (*Hearings*, p. 1412). That is to say, it was assumed that the applicators would wear protective clothing, observe all the handling and cleanup instructions on the label, not be subjected to accidental spills, and so on.

Monsanto held that the government not only had overstated the risks of alachlor, it had grossly understated the risks of its chief alternative, metolachlor. Monsanto produced confidential studies of other chloracetanilide compounds which, according to the Review Board *Report*, gave evidence of tumour reactions similar to those produced by alachlor. Even though the lowest dose levels of metolachlor that produced oncogenic or carcinogenic response in rats were significantly higher than the lowest dose of alachlor to do so, Monsanto argued that the differences in the studies (different strain of rats, different experimental protocol) meant that the only firm conclusion to be drawn from these studies was that metolachlor, like alachlor, was an animal carcinogen. It could not be concluded, however, that it was a *less potent* carcinogen than alachlor (*Hearings*, pp. 5218ff.).

Monsanto also pointed out that water studies done in areas where both herbicides were used revealed that metolachlor was present in as high levels as alachlor and, the company claimed, it stays in the

environment longer. In Monsanto's opinion this undermined the
government's claim that metolachlor was a safer, or even a completely safe, alternative to alachlor.

Nevertheless, Monsanto also argued that the finding of alachlor in
wells and municipal water supplies did not constitute a serious risk
to the public. The studies did not prove that alachlor was leaching
into the ground water, but only that surface run-off water was getting into poorly constructed wells. The underlying, but largely
unstated, assumption in Monsanto's argument was that whatever
risk was posed by the levels of alachlor present in wells was a risk
that was being taken *voluntarily* by the users (since they are responsible for the poor construction), and hence was not being *imposed* on
them by Monsanto's product. Further, Monsanto pointed out that
the levels of alachlor found in the wells was far enough below the
levels of known carcinogenicity in rats to represent a clear margin of
safety. As for the alachlor present in the municipal water systems,
Monsanto argued that it could be removed by the installation of
charcoal filtration systems.

Monsanto's position on the matter of safety was that Risk-Benefit
Analysis was the best way for regulatory agencies to approach the
matter. The company even argued that the PCPA *required* the government to adopt this approach (*Report*, p. 26). Thus, throughout the
debate Monsanto expressed the view that the safety of alachlor could
be determined only by weighing the risks it posed against the benefits obtained from its use. The company's view was that whatever
risks are potentially posed by alachlor, the clear benefits of its use far
outweigh them.

In Monsanto's view, the government's primary error was that it
did not consider these benefits in its cancellation decision, but
attempted to rely upon some "absolute" standard of acceptable risk
which it never clearly articulated. Monsanto counsel and witnesses
argued that such "absolute" standards, which do not weigh risks
against benefits, are "arbitrary and capricious" because there is no
good reason to set them at any one point rather than another (*Hearings*, p. 5221ff.). Weighing risks against benefits was the only non-
arbitrary standard. What is an unacceptable level of risk in one case
may be acceptable in another if the benefits are greater.

As Monsanto counsel put it, alachlor is not like Red Dye No. 3,
where "all it does is colour lollipops and if there is a remotest suggestion that it gives cancer in the tail of a rat, we're not going to use
it." It is more like the case where, if there is "a chemical that will
cure 10 per cent of the people with Parkinson's disease, but it does
other terrible things, we're going to allow it on the market" (*Hearings*, p. 5243). Clearly, Monsanto believed that the value to Canadian

farmers of using herbicides is more akin to the value of saving human lives than to the value of the increased pleasures derived from lusciously coloured lollipops. Several Monsanto witnesses put the case even more strongly, claiming that chemical weed control is a "necessity," not merely a supplementary option. It is "necessary" in order to compete on equal terms in the world agricultural market. This "necessity," of course, is created by the fact that others in the market (e.g., the American farmers) are permitted to use alachlor.

Although Monsanto argued that Risk-Benefit Analysis was the proper approach to safety questions, it also contended that alachlor could meet even the more rigid test of "No measurable human health risk," or the NOEL standard (*Hearings*, p. 121). Given its interpretation of the rat studies, there was no evidence of animal or human health risk below those levels of expected human exposure. Therefore, strictly speaking, there was no "risk" to be balanced against the substantial benefits of alachlor use. At worst, there was only the potential for unexpected and improbable exposures above the hazardous levels in animals to be weighed against the benefits.

In the final analysis, however, Monsanto's argument came down to the question of regulatory fairness. Its major complaint against the government was that it had acted to ban the company's product from the market without sufficient evidence. The company believed that the best way to do risk estimation was to use rigorous, quantitative methods. However, the data available to the government (which, of course, it had depended upon Monsanto to provide) were not complete enough to permit a rigorous quantitative assessment. In the absence, then, of any *reliable* data, the government had no non-arbitrary basis for declaring alachlor unacceptably risky. In other words, Monsanto believed that the onus was on the government to prove with reliable, quantitative data that alachlor was unsafe, not on the company to prove with its own studies that it was safe. As we shall see, this turns out to be one of the most significant points of difference in this controversy, and we suspect in many risk debates.

Summary of the Alachlor Review Board's Findings

In its published *Report*, the Board concluded, on the basis of its own estimation of the risk of alachlor, that the Minister's decision to cancel the registration of alachlor was unwarranted. It is important to note that the Review Board interpreted its mandate as not limited simply to reviewing the data available to the government, to determine whether the government had acted judiciously. Rather, it undertook to do its *own* risk estimation, admitting to its investigation many additional studies, and the testimony of many expert witnesses, data to which the government had no access when it issued its cancellation order.

On the basis of its own independent risk estimation the Board recommended that the Minister restore the registration of alachlor. Its decision was reached on the basis of the following conclusions and considerations concerning the safety of alachlor.

First, the Review Board agreed with the government that alachlor is an animal carcinogen, and that it "should be considered to be a *potential* human carcinogen for regulatory purposes" (*Report*, p. 7). It accepted the evidence of the feeding studies with rats, but disagreed with the government's view of the mouse feeding studies and held that the data permitted the conclusion that alachlor causes cancer in only one animal species, not two, as the government had claimed.

The Review Board rejected the government's occasional characterization of alachlor as a "probable" human carcinogen, substituting the term "potential," on the grounds that the government's term was too strong, carrying the implication that there was a greater than 50% probability that alachlor caused cancer in humans. In the Board's view, the data from all the chronic feeding, metabolism, and epidemiological studies provided no basis for any conclusions about the probabilities of alachlor's carcinogenicity.

Secondly, the Review Board disagreed with the government about the safety of metolachlor, the primary substitute for alachlor, and adopted Monsanto's view that metolachlor also was an animal carcinogen and should be considered a potential human carcinogen. In addition the Board agreed with Monsanto that even though much higher doses of metolachlor than of alachlor had been required to produce cancerous responses in the animal studies, the data were not clear enough to draw a "scientifically" reliable estimate of any difference in the carcinogenicity of the two chemicals. Thus, for regulatory purposes they should be considered to pose equal risks in rats and equal risks in humans.

This conclusion was critical to the Board's decision to recommend the reinstatement of alachlor. It undercut the government's implicit assumption that there was a safer alternative to alachlor on the market, so there were no benefits of increased safety derived from cancellation that could be put into a balance against the significant adverse consequences to farmers. In the Board's view, if there was good enough reason to cancel the registration of alachlor, there was equal reason to cancel metolachlor. However, the economic consequences of such action would outweigh the potential risks posed by allowing both chemicals on the market. The Review Board agreed with Monsanto and the farm representatives who testified before them that the continued registration of one or the other of these two herbicides is "essential" if corn and soybean production in Canada is to remain economically viable and internationally competitive.

Further, the Review Board found that if either of the two herbicides were removed from the market, there would be adverse economic impacts, particularly upon farmers, as a result of the monopoly situation created. This would be felt primarily in the rise in the price of metolachlor, produced by Monsanto's chief competitor in the agricultural herbicide market, Ciba-Geigy, Inc., despite the latter's assurances to the Board that it would not raise the price in the monopoly situation, and had not done so in the period since the cancellation of alachlor during which it enjoyed a monopoly.

On the question of the levels of exposure to alachlor by applicators, the Review Board concluded that the government's estimates had been far too high and were based on an "unreasonable worst-case" scenario. The Board found the "reasonable worst-case" estimates of applicator exposure to alachlor to be 1,000 to 10,000 times (three to four orders of magnitude) lower than the lowest dose at which a tumour was observed in the long-term rat feeding studies. The Board considered this to be a "reasonable margin of safety." Here the Board found itself in a position midway between the government's and Monsanto's estimates. It disagreed with the government's decision not to amortize the exposure at all, but also with Monsanto's decision to amortize over a full lifetime. It also rejected the government's reliance upon skin patch studies which assumed 100% absorption of the amount of alachlor deposited on the skin into the bloodstream, and gave more credence to Monsanto's biomonitoring studies of alachlor levels in urine which suggested much lower levels of exposure. The Board also believed it was reasonable to assume that the protective clothing recommended by the labelling on the product would be worn by the applicators, an assumption the government had thought unreasonable. All of these assumptions led the Board to arrive at exposure estimates dramatically lower than the government's, and low enough to constitute a "reasonable margin of safety." We show in the following chapters that these critical assumptions in the risk estimation of alachlor were founded on important value considerations.

With respect to the drinking water contamination problems identified by the government, the Board concluded that "reasonable worst-case" estimates of potential public exposure are even lower than the applicator exposure estimates, and this, too, represented a "reasonable margin of safety."

The Review Board agreed with Monsanto that there is no reason to estimate public or applicator exposure to the alternative herbicide metolachlor any lower than to alachlor. Given that the likely human exposure levels from the two chemicals are similar and that the carcinogenic potency of the two are also similar, there is no basis for the government's cancellation of the one herbicide while leaving the other on the market. And since the economic costs of cancelling both products far outweighed the slight risks assessed in the "reasonable use" of the herbicides, the Review Board concluded that both should be allowed on the market.

The Board chided the Minister for not paying attention to the issue of the benefits that are derived from the use of these chemical herbicides (which, in the Board's eyes, were clearly discernible and substantial) and putting them into the balance against the risks (which it judged to be less clearly discernible and insubstantial). It is not appropriate to consider "safety alone" as the government had claimed to do. The Board admitted that the risk-benefit approach was not required by the PCPA but, nevertheless, it was the approach the Board thought most reasonable for regulatory bodies to follow. All other "absolute standards" it considered to be arbitrary and unclear in their application.

Other Parties in the Alachlor Review Board Hearings

In addition to the three major parties to the alachlor dispute whose positions we have just summarized, others made submissions to the Review Board Hearings, and participated in the oral testimony and in the cross-examination of witnesses. Among these were representatives of Ciba-Geigy and representatives of various farm and environmentalist organizations. Among the environmentalists was a lawyer, Toby Vigod, who represented a farm-wife, Mrs. Zelma Van Engelen, from Lambton County, Ontario. Mrs. Van Engelen had special interest in the alachlor debate due to the fact that alachlor (and other farm chemicals) had been discovered in her well by Ontario Ministry of Environment officials at levels considered dangerous by the federal government, during the time that she was breast-feeding her baby. Other environmentalists represented at the Hearings included the Friends of the Earth and Pollution Probe.

Each of these various additional parties to the alachlor proceedings represented clearly discernible interests, or "value frameworks" (as we define this in the following chapter). These may have been more clearly evident in the positions they took in the controversy than the value frameworks of the other three major parties, though we believe they were no less influential. The major interest of Ciba-Geigy, of course, was to counter the claims of Monsanto that the former's product metolachlor posed as great a risk to human health as alachlor. Because of this interest, Ciba-Geigy tended to side with the government's position, especially its claim that metolachlor was less risky than alachlor, as well as its position that alachlor's risks should be evaluated independently of any comparison with other products.

The representatives of the farm organizations clearly were most concerned with the detrimental economic consequences for their members that might result from the banning of alachlor and other chemical herbicides. While they also indicated that they were concerned about the health risks posed to farmers by these products, this did not come through in their testimony as a primary concern.

The dominant value represented by the environmentalists, of course, was concern with the detrimental effects of chemical herbicides on human health and on the ecology of farm regions. In consequence, they strongly supported the government's view that the potential margins of error in the estimation of alachlor's risks should be interpreted on the side of safety rather than economic benefits. They also were the most clearly opposed to the use of risk-benefit approaches to the question of risk acceptability, insisting that the government was permitted by the legislation to ban a herbicide on the showing of risk alone, independent of any balancing against offsetting benefits. Their representatives, particularly Toby Vigod and J. F. Castrilli, played a significant role in the Hearings, especially as cross-examiners of other witnesses. While they did not attempt to present to the Review Board a comprehensive risk estimation of their own, they raised many important questions about the uncertainties that plagued the studies presented to the Review Board by others, and identified the assumptions underlying many of the interpretations of the data.

One of the most fascinating features of the Hearings, evident to anyone reading through the transcripts, is the attitude of the other participants, particularly the Review Board members, toward the environmentalists. They seemed to be perceived as something of an annoyance, or at least as not properly a part of the proceedings. This may have been due in part to the fact that they represented a value framework perceived as antithetical to that of the Board members. But, in our view it was also due to the fact that, because the value commitments of the environmentalists were more salient their intervention was viewed as inappropriate in what was thought to be a "scientific" inquiry.

Overview

In the following chapters we analyze the arguments of each of the major actors in the alachlor debate outlined above. We are interested in uncovering the factors that produced the wide discrepancies in the conclusions the parties reached, both at the level of their *estimation* of alachlor's risks (as well as metolachlor's), and at the level of their judgement about the *acceptability* of these risks.

Before jumping into that analysis, in Chapter II we first look at the way in which the traditional, or classical, model of risk assessment understands its methodology and explains its discrepancies when used as a regulatory or mandated science. We then sketch out an alternative model of risk assessment that we believe more accurately describes the way it actually functions, particularly in the role of regulatory or mandated science. We suggest that these assessments are "framed" by a definable and coherent value framework that significantly influences critical choices made in the interpretation of data along the way. The greater the uncertainties in the data, the more influential this value framework becomes in the risk assessment.

One of the places where these values make their most significant impact in the regulatory situation is in the decision regarding the assignment of the burden of proof, to either the case for safety or the case for risk. This critical decision we believe to be at the heart of the debate about alachlor. A second decision closely related to the first, concerns whether to hold the risk estimation to the standards of rigorous, laboratory science. And a third decision relates to the attitude one takes toward risk itself.

It is our view that the value framework which framed the Alachlor Review Board's choices in each of these cases and, hence, its overall estimation of alachlor's risks, was founded upon three clearly identifiable normative premises. The first was a classically "liberal" conception of society and the role of government. The second was a positive view of the role of technology in human society. And the third was an instrumental conception of the nature of rationality.[10]

In the following chapters we test our alternative model and our hypotheses about the norms at work in the alachlor debate against the actual arguments put forward by the major parties in their estimations of the herbicide's risks and the acceptability of those risks. In Chapter III we probe the arguments of the government and Monsanto. In Chapter IV we scrutinize the position of the Review Board as it is argued in the *Report*. In Chapter V we focus specifically on the choice of the risk-benefit standard of safety adopted by Monsanto and the Board, and explore both how this standard is motivated by their value framework and how the standard tends to feed back into, and influence, the risk estimation. We end our discussion in Chapter VI with a closer examination of the value framework that appears to undergird the Review Board's risk estimation. We are interested in how the various elements of this framework reinforce and complement one another and how they all relate to an identifiable attitude toward risk and risk-taking itself.

Our conclusion is not that these value judgements in any way undermine or discredit the Review Board's conclusions. Indeed, we hold that these value judgements are an inevitable part of risk estimation as a regulatory or mandated science. Neither do we necessarily contest the particular value framework underlying the Board's conclusions. We merely point out that the different estimates reached by the government, and by other parties who believed with the government that alachlor's risks were not acceptable, do not necessarily stem from the fact that they conducted their investigation worse or better than Monsanto or the Review Board, but rather that they did not share the same value framework. If this is true, then it appears that more is at stake in risk debates than can be settled with science alone.[11]

II

An Alternative Model of Risk Assessment

In this and the following two chapters we will review in some detail the risk estimations conducted by the Crown, Monsanto, and the Alachlor Review Board. Our primary finding is that these estimations were decisively guided by differing value frameworks maintained for the most part implicitly and without recognition by the estimators. These value frameworks differentially framed the risk assessment process for each of the three major parties to the alachlor debate. Moreover, as we shall show, the decisive influence of values on the conclusions concerning alachlor's risks reached by these three parties to the debate did not mark an aberration or failing in their estimations, a sign of avoidable "subjectivity" skewing what otherwise could and should have been objective and value-free assessments.[12] Rather, the structure of the risk assessment project was such that this influence of values was unavoidable and unavoidably decisive. That is, the process of estimating the risks of alachlor was essentially *political*. This conclusion, which we attempt to substantiate in the following pages, would not be particularly challenging if we were speaking of risk *management*; but the process under consideration in this and the following two chapters is risk *estimation*.

Finding risk estimation to be an essentially political process marks a radical break with the classical model. The classical model recognizes that different risk estimates of the same project may vary considerably (those for alachlor being a dramatic case in point), and that it is scarcely possible in practice for different assessors, however "objective" they may be, to reach consensus. Nevertheless, the idea that behind the diverging estimates there is one "objective fact," the *actual* risk, which the various assessors are attempting to identify, and the conviction that consensus is a theoretical possibility, provide a powerful guiding image. Any analysis of competing risk estimations is set into a value-neutral context, focusing issues on gaps and errors in the data base. The implicit assumption is that, with enough information, all competing estimates would converge. This view of risk estimation as value-neutral is reinforced by the tendency in practice to eliminate controversy from the initial hazard identification by choosing for study limited hazards out of the entire spectrum of hazards, especially those that are easily quantifiable, such as the risk of loss of life or "life years." The classical model thus provides an analytic framework within which the process of *estimating* risk is pictured as separable from the public debates about *managing* risk.

This separation does not hold up under our analysis of the risk estimations performed by the government, Monsanto and the Alachlor Review Board. As a result, the model itself explains little of these processes. Its primary assumption that the factual can be separated from the normative, the descriptive from the prescriptive; its associated goal of providing a value-neutral risk estimate; and its focus on the theoretically objective, calculated outcome, run counter to the realities of the risk assessment process, which is dominated by human judgement in the face of uncertainty. Because the model does not acknowledge the role of value-based judgement in risk estimation, it is unable to account for the dual roles of science and policy that comprise that judgement. It cannot identify the social values brought into play by the public policy aspect of the risk estimate, which produce competing interpretations of uncertainties in the data base. And it does not provide a framework for the assessment or critique of these values.

A more realistic model of risk assessment, one that is sensitive to the role of values in the estimation of risk, is urgently needed. In this chapter, we sketch the main outlines of such a model, as suggested by our analysis of the alachlor debate. Thus, the chapter is largely anticipatory. As indicated above, the risk assessor's value framework contributes to his or her manner of *framing* the risk assessment process. This image of "framing" appeals to the obvious fact that in the context of risk estimation, scientific data do not interpret themselves; to determine what the data indicate concerning the risks of a product, the assessor has no alternative but to employ an interpretive point of view.

Defenders of the classical model of risk assessment will not feel threatened by the suggestion that risk estimates are guided by interpretive points of view. Rather, the challenge to the model results from the fact that, on our account, normative assumptions are indispensable ingredients of these points of view.

While we believe that the ideal of the classical model—of risk assessments built upon rigorous science—simply does not describe what happens in these assessments, nor even what could or ought to happen, it does not follow that we deprecate science or minimize its role. Our model of risk assessment simply acknowledges the interconnections between the scientific and social policy elements. The quality of the scientific element is a separate issue. If a social objective constricts or biases the scientific component of a risk assessment in such a way that scientific problems arise—valid data cannot be produced, invalid data are treated as valid, or scientifically unsupportable inferences are made—then that social objective can be argued to have subverted the risk assessment. Such an assessment could not be seen as valid. Social objectives must operate within the bounds of the scientifically credible, since bad science yields not only bad risk assessments but bad social policy. Our alternative model of risk assessment concurs with the classical model on this point.

If the assessment is undertaken within a specific regulatory context, as was the case with the risk estimates of alachlor presented during the Alachlor Review Board Hearings, then important elements of the interpretive point of view, or, as we shall refer to it, *frame*, will be given by that context. In the Hearings, the regulatory context very largely settled the sorts of issues that would be relevant to the investigation. For example, it defined the overall assessment project as one concerning the risks of alachlor, identified cancer as the significant hazard, and established that it would be irrelevant to consider the risks of alachlor in isolation from the risks of metolachlor. These are but some of the more obvious ways the assessment process was framed owing to its placement in a distinctive regulatory context. In a more elaborate account of the model we are sketching out here, it would be necessary to point as well to wider institutional settings behind the regulatory context, which similarly contribute to the specification of those issues relevant in the risk assessment process.

Thus, the regulatory and wider institutional contexts of a risk assessment fix the issues considered relevant for the risk assessment, the issues that need to be responded to so that the product's risks might be properly estimated. In this way, the regulatory and wider institutional contexts set a frame within which the risk assessment will be carried out. In addition, of course, the scientific investigation, which is guided by this specification of issues, generates its own issues. For example, in the alachlor debate the regulatory context dictated the focus on cancer as the relevant hazard to be investigated.

But the rat feeding studies, undertaken in order to estimate the carcinogenic effects of alachlor, raised further questions, such as
whether different strains of rats metabolize alachlor differently.

Our review of the alachlor debate reveals an unexpected feature of
these issues—unexpected, at least, from the perspective of the classical view of risk assessment. This is that many of the significant issues
fall into one or another of two classes. Some are of a kind that we
shall refer to as "inherently normative"; others, of a kind that we
shall refer to as "conditionally normative." Our suspicion is that
both sorts of normative issue are present, and unavoidable, in *all* risk
assessments. In calling these issues "normative," we are contrasting
them with other straightforwardly empirical or scientific issues.[13] A
normative issue is one that can only be resolved by invoking values,
views concerning fairness or justice, or views concerning what
"ought to be" as opposed to "what is." Whereas responses to normative issues are "prescriptive," responses to empirical issues are
"descriptive." In the alachlor controversy, the normative issues
were decisive for the estimation of alachlor's risks. But, as we shall
see, they are not issues of a sort that a value-free science can resolve.

It is this feature of the controlling issues in the risk assessment of
alachlor that makes the classical view of risk assessment an inappropriate model of the alachlor debate. We shall begin our sketch of the
alternative model of risk assessment by identifying some of the
major normative issues with which the various risk assessors of
alachlor grappled.

Inherently Normative Issues

The idea of "inherently normative" issues is clarified in Chapter IV.
But to anticipate that discussion, so that the significance of this component of the alternative model might be appreciated, we list here
some of the inherently normative issues that Monsanto, HPB, and
the Review Board needed to respond to in the course of assessing the
risks of alachlor:

1. Is it fair to expect that alachlor applicators will wear protective
   clothing, including especially adequate gloves?
2. Is it fair to expect that commercial applicators of alachlor will use
   closed cab tractors and closed application systems?
3. Is it fair to include in estimates of applicator exposure to alachlor,
   exposure that results from spills or from other accidents or results
   of carelessness?
4. Is it fair, when estimating alachlor exposure that results from contaminated well water, to include exposure caused by poorly constructed wells?

5. When is an exposure scenario a "reasonable worst-case" scenario?

It is evident that none of these questions is purely, or even primarily, empirical; all are inherently normative. In saying this, we are pointing to the obvious fact that each invites a response concerning what *ought to* be the case, rather than what *is* the case. Thus, in answering each question one is *prescribing*, not *describing*. This is not to say, of course, that facts are irrelevant to resolving issues of these sorts, but is only to identify the nature of the issue resolved.

As will be seen in Chapter IV, these issues proved to be crucial in the estimation of alachlor's risks. The assessors' final estimates were far more sensitive to differing responses to these issues than to differences concerning the outright empirical issues that were also addressed. But what motivates and makes reasonable any particular set of answers to these inherently normative questions is the value framework held by the risk assessor. (Of course, a final judgement concerning the reasonableness of such answers awaits a decision concerning the reasonableness of that value framework.)

It may seem surprising that normative issues of the sorts identified should have loomed so large in a risk assessment process that was represented by all major parties as strictly objective and scientific. All *could* have been defined as empirical rather than normative issues. For example, the Board could have understood the question about protective clothing as inviting a prediction. *Will* applicators wear protective clothing? rather than, Is it fair to expect them to do so? Even the fifth question theoretically could have been framed empirically—as a simple prediction of the highest levels of exposure possible. In actuality it is not empirical in this way because it is a question about what level of exposure *ought* to be taken seriously. The answer to this question, of course, will be tied to responses to several of the first four normative issues.

It is important to see why these issues were defined as they were, else one might conclude that presence of the inherently normative issues represents an avoidable aberration in the alachlor debate, and therefore of no general significance. The critical fact is that the risk assessment took place in a regulatory context. The assessors were making an input to a decision that would affect the rights of various interested parties as well as the well-being of the public. To ignore this fact, by formulating the issues in an empirical way (so that, for example, the first issue would have become whether applicators actually wear protective clothing), would not magically cause the normative significance of the issue about clothing to disappear. That is, *if* it would be unfair to Monsanto to cancel alachlor's registration because of the exposure incurred by unreasonably careless applica-

tors, then the risk assessment *should not* take account of such exposure. *That* issue would not go away just because a determinedly
"empirical" assessor chose to redefine the inherently normative
issues in the risk assessment as empirical or predictive ones. Thus, it
is to the Board's credit that it properly construed the inherently normative issues, even though, as suggested, it did not recognize what it
was doing.

The inevitability of inherently normative issues in risk assessments
may be seen by reflecting on the first three issues identified above,
those concerning protective clothing, the sort of tractors used by
applicators, and the relevance of spills. The context is that the risk
assessor is trying to estimate exposure. But the exposure conditions
vary. For example, some applicators wear adequate protective clothing, some do not. In these circumstances, to estimate exposure it is
necessary to *normalize* the exposure conditions by choosing from
among the various practices actually followed by applicators those
that should be assumed for purposes of making the estimate. But the
question, How shall we normalize the exposure conditions? is normative and not empirical, prescriptive not descriptive. That is, in
answering it, one inevitably stakes out a position concerning values.
To normalize the exposure conditions by assuming, for purposes of
the exposure estimate, that applicators wear adequate protective
clothing is to accept the idea suggested in the preceding paragraph:
fairness to the manufacturer dictates ignoring the exposure incurred
by farmers who do not follow the instructions on container labels. To
normalize exposure conditions in the other direction, by assuming
that applicators do not wear adequate protective clothing, is to reject
the fairness argument in favour of the value position that what matters most is farmers' health.[14]

Conditionally Normative Issues

Other issues, which we call "conditionally normative," arose
because of the significant presence of uncertainties in the information on which the risk assessment was based. They are issues which
theoretically or ideally are purely descriptive: they could be settled
by purely empirical means *if there were enough empirical data*. They
are "conditional" in the sense that they come into play only in the
condition where there is not enough information available to decide
the issue without appeal to them. Their normative character derives
from the context of uncertainty in which they are situated, owing to
which a mere appeal to the "facts of the case" is insufficient to
resolve them. Instead, given the fact that the necessary empirical
information is not available, they are issues that can only be resolved

by invoking values—ultimately, the value framework held by the risk assessor. Our point is not merely that these issues are resolved by invoking values, but, more, that where the condition of uncertainty obtains it is reasonable and necessary to resolve them in this way.

In the alachlor debate, most of the issues confronted were conditionally normative. We cite here just a few prominent examples:

(A) The effect of the stabilizer used in commercial alachlor on its oncogenicity, including possible synergistic effects (*Report*, pp. 52, 63).

(B) The most significant route of exposure, particularly in the nasal turbinates. Nasal turbinate and stomach tumours could be induced by local irritation; liver tumours on the other hand must be induced systemically by material absorbed and carried in the blood stream (*Report*, p. 63).

(C) The minimum exposure period necessary to produce tumours in rats, and the onset of the tumours found in each study. They were killed after two years; interim kills were not carried out (*Report*, p. 62).

(D) The relative importance of adenoma (benign glandular tumours) formation in adenocarcinoma (cancer) induction (*Report*, pp. 63, 64).

(E) The existence and significance of evidence of genetic mutation.

(F) The relative importance of the long-term (two-year) studies on carcinogenicity versus the short-term genotoxicity studies.

(G) The significance of the brain tumours which were found in later analysis to originate in the nasal area (*Report*, p. 55).

(H) The use of concurrent versus historic controls in the calculation of statistical significance.

(I) Whether patch tests or biomonitoring is the appropriate method of determining applicator exposure.

(J) Whether exposure should be amortized, and if so at what rate.

The principal complexity in our alternative model of risk assessment results from the subtle ways in which values and, ultimately, value frameworks, control the answers to the foregoing questions and others like them. We offer below an admittedly incomplete account, which focuses on the ways uncertainties in the data base for the assessment of alachlor were circumvented, so that despite indeterminate data determinate risk estimations could be made.

We begin with an anticipatory explanation of the idea of a value framework, follow that with a general discussion of the role of uncertainty, and continue with a more analytical account which suggests how, in the face of uncertainty, risk assessors make their way by adopting a variety of "argumentative strategies." These strategies form responses to the following questions: (1) How rigorous should the science on which the risk estimation relies be? (2) Who has the burden of proof? (3) What attitude toward risk should be adopted, that of a risk-taker, or that of one who is risk-aversive? Their position on these matters completes the "frame" by which they interpret the scientific data available for assessment. The risk assessor's value framework "intrudes" into the risk assessment process owing to the necessity of appealing to values in support of a position concerning each of these three critical matters.

Value Frameworks

We use the term "value framework" to indicate that the values influencing a risk estimation are not a haphazard collection but form a pattern. Hence, the idea of a "framework." No doubt different risk assessment projects will determine different sets of issues to be relevant, although we expect there will be a common central core. In the case of the alachlor debate, the central issues refer to the role of government *vis-à-vis* the private sector, the place of technology in our lives, and the nature of rationality. Of principal interest below will be the positions of the various parties, especially the Review Board, regarding these three issues, as this can be inferred from their statements and deliberations.[15] For example, with respect to the first issue the Board took a "classical liberal" view of the role of government; with respect to the second, it took a positive attitude toward technology; with respect to the third, it adhered to that conception of rationality often referred to as "instrumental rationality."

Because they are interrelated, views on these three issues typically fall into distinctive patterns; many of the logically possible combinations of responses to them are unlikely to be met, and many such combinations are, in a sense, internally inconsistent. Responding to the first and third of the issues in the way the Board did, while being a neo-Luddite with respect to the second issue, would illustrate the sort of inconsistency we have in mind.

It can be argued that a value framework is, at the personal level, a commitment: one cannot readily dispose of it. Instead, it forms a basic part of his or her identity. In a fuller statement of the alternative model than that attempted here, it would be important to explore the relations between personal identity and value frameworks and to consider the sources of the characteristic identity underlying contemporary risk assessment practices. Without that elaboration, however, we can see that owing to the connection between one's value framework and one's manner of identifying oneself, a technology debate is likely to be a much more *personal* confrontation than is generally recognized.[16]

Views on the three indicated issues are complemented by more specific positions regarding values. These form the channels, as it were, by which the assessor's more general value framework makes a coherent risk assessment possible. To simplify a complex matter, in the alachlor debate the principal value at work in the government's risk estimation was human health; by contrast, the Board gave priority to economic benefits. To illustrate the connection between these specific values and positions taken on the more general issues that define an assessor's value framework: the Board's concern to protect the competitive position of Canadian farmers exhibited the sort of willingness to trade off costs (risks to human health) for benefits (farm income) that characterizes instrumental rationality, and its concern for the freedom of Monsanto to market its product was a rather straightforward expression of classical political "liberalism." In view of these connections between dominant specific values and positions taken on the three issues, we may represent the risk assessor's relevant value framework as layered, with positions on the three issues (e.g., the role of government, the place of technology, and rationality) forming the deep layer, and views concerning the ranking of specific values (e.g., health versus income) the surface layer.

Uncertainty and the Role of Judgement

Risk assessment, as an activity in the domain of regulatory science, is practised in a context of uncertainty. We are using the term in a broad sense to refer to issues that cannot be convincingly settled by appeal to what is scientifically known or issues that divide the scientists working in the area. Thus, there may be uncertainty concerning the data, their interpretation, and the conclusions to be drawn from them; as well, there may be uncertainty concerning the appropriate methods to follow. The significance of uncertainty is that at every turn the assessor faces questions to which no unequivocal answer can be given. The facts do not speak loudly and clearly. But because risk assessment is an applied science, risk assessors do not have the luxury of responding to the uncertainty by frankly admitting that they don't know the answer and so must simply remain silent. The essential feature of applied science in a context of uncertainty is that judgements must be made despite there being no conclusive evidence at hand for making them.

Two sorts of uncertainty are especially evident in risk assessments. These are sometimes referred to as ontological and epistemological uncertainty. By ontological uncertainty is meant uncertainties *in the world*. A prediction is needed concerning an event which, at the time, is undetermined: it might or might not occur. Examples include many (but not all) of the events that fall under the heading of "operator error." The difficulty is not that one lacks sufficient information to decide whether a problem will develop in the operation of a nuclear power plant, say; rather, it is not yet determined in reality whether the operator will err and thus create the problem.

Epistemological uncertainty, on the other hand, refers to uncertainties arising from limitations *in the data base*, which lead to either information gaps or ambiguity, or from lack of a scientific basis for resolving methodological disputes. All of the examples of conditionally normative issues cited above result from epistemological uncertainty. In principle, uncertainty of this sort can at least be reduced, if not eliminated. But for the risk assessor, working under time constraints and faced with the necessity to reach a conclusion, the abstract possibility of reducing the uncertainty yields scant comfort. And in many areas the further research that is designed to overcome gaps in the data, eliminate ambiguity or vagueness, and otherwise reduce uncertainty only serves to open up new domains, with yet new uncertainties.[17] In the alachlor review process, the only statistically certain relationship available to data analysts was that between alachlor and cancer in rats, at doses well above the range of expected human exposure. Most other conclusions related to the exposure, absorption, metabolism and carcinogenicity of alachlor in humans required a great deal of discretionary judgement.

The reflection which, above all, explains the relevance of uncertainty for risk assessment is the following: In making judgements on the basis of inconclusive data, analysts are naturally motivated to protect whatever they value most. Indeed, as we shall show, they have no alternative to doing so. If health is the most valued consideration, for example, while economic benefits are regarded as more expendable, then the analyst is more likely to judge each uncertainty in light of the potential implications for health, assessing the risk at the high-risk end of the uncertainty continuum, rather than in the light of potential benefits to be derived from an estimate at the low-risk end of this continuum.

An analogy will suggest the inevitability of proceeding in this way. Imagine a golfer preparing to hit her tee shot down a fairway that is bordered on one side by a rough with mostly good lies and on the other side by a bunker. The fairway between is narrow and it is uncertain whether the golfer can manage to keep her drive on the fairway. If the golfer is weak on bunker shots, she may decide to aim her tee shot down the rough side of the fairway, realizing that the bunker poses more danger to her than the rough and thus preferring the rough to the bunker, in the event she should miss the fairway. If, by contrast, she sees herself as a strong bunker player, but weak out of the rough, she will likely aim her tee shot down the bunker side of the fairway. Golfers typically react in this way, and it is rational for them to do so.

The similarities are obvious: risk assessors who value human health, which is threatened by a product, above economic benefits to be gained by marketing the product, will respond to uncertainties concerning the extent of the threat (and of the benefits) by giving more credence to the possibility of higher risk than they would if they valued the potential economic benefits more than human health. In effect, they will aim for the rough side of the fairway rather than the bunker side. The analogy points up an important feature of the case. Just as the golfer does not *aim* for the rough, but only for the rough side of the fairway, so risk assessors do not deliberately over-estimate the risk. The attempt is to estimate the risk correctly, just as the golfer hopes to end up in the middle of the fairway; only this is qualified by the consideration that *if* they are going to err they prefer to err on the side of an over-, rather than an under-estimation.

The analogy suggests an objection to our claim that when there is uncertainty the risk assessor has no alternative to invoking values. In golf, it might be argued, it is better to forget both the rough and the bunker and aim the shot exactly where the golfer most wants it to be. The greater one's confidence in one's game, of course, the more reasonable it is to adopt this stance, since the risks of ending up in either the bunker or the rough decrease as one's skill increases. Similarly, it might be argued, the risk estimator should simply concentrate on getting the best estimate, forgetting the values that are at stake in the event of an over- or under-estimation. Rather than attempt to compensate for possible error by moving one's risk estimate out to the margins of error, one should, in effect, aim "down the middle of the data."

If there were no reply to this objection, our thesis concerning the inevitability of values in risk estimations would be somewhat weakened. It would remain the case that values are involved in resolving the *inherently* normative issues faced in risk estimations. And it would remain the case that where there is *extreme* uncertainty there is no discernible "middle" of the data to aim down. (As will be noted below, much of the uncertainty confronting the alachlor assessors was of this extreme sort.)

But what shall we say about the case where the uncertainty is not extreme, so that the scientific dimension of the risk estimate, taken by itself, affords some indication concerning the position the assessor should take? It seems evident that, even when the uncertainty is low, the decision to proceed by "aiming down the middle of the data" cannot be taken without making a value judgement. The reason is that to the extent there is uncertainty there is also the possibility that one might err on one side or the other—on the side of over- or under-estimation of risk. There is, then, an implicit willingness to accept these risks of error in order to realize the benefits of "being right" in one's assessment. If the "facts-only" estimate produces the conclusion that there is no risk, or that the risk is "acceptable" by some standard, while the uncertainties are such that the risk *could* turn out to be unacceptably high, then the decision to "go with the facts alone" would be a *risk-taking* decision. As such it would depend on being more impressed by the benefits to be gained in the event the gamble succeeded than by the losses that would be sustained, in the event it failed. Judgements of this sort are influenced by the degree of one's confidence that the gamble will succeed. That is, the risk assessor who recommends value-free estimates in a context of mild uncertainty may be recommending a risk-taking approach, and the decision between a risk-taking and a risk-aversive approach expresses a value judgement.

On the other hand, if the "facts-only" estimate produced a conclusion that there *was* a risk to be concerned about, while the uncertainties of the data suggested a margin of error that embraced the possibility of no risk or acceptable risk, then to ignore that possibility would be to take a *risk-aversive* stance. The greater the uncertainty, the wider the margin of error, and the greater the possibility that the estimate might be wrong and that one is foregoing possible benefits in being deterred by the "facts only" estimation of risk. So the point is not that the attempt to attend only to "the facts of the case" is necessarily risk-taking (or risk-aversive), but rather that, depending on the circumstances, it will reflect one of these values.

The point is of considerable importance to our argument. Consider again our golfer. Suppose she fears the bunker but decides nevertheless to aim the ball to the bunker side of the fairway, since that is the optimal position from which to make the second shot. Unless she is *absolutely* certain that she can aim for that position without ending in the bunker, she is taking the risk of finding herself in that predicament. Her decision to "play it straight" is a risk-taking decision. The less capable she is of hitting her target, the greater the risk she takes. It does not matter what her *reason* is for making this choice. It might be that she deems the expected gain of making a good shot to outweigh the risk of landing in the bunker. It might be that she simply overestimates her abilities. Or perhaps she simply follows the rule,

"Always aim directly for the optimal position, regardless of hazards." Regardless of which of these is true, it is still the case that hers is a risk-taking, and therefore a value-laden, decision.

Similarly, risk assessors may be determinedly "scientific," and never entertain the idea of looking to any aspect of an estimate beyond the purely factual. They strive to be "neutral and value-free." Thus, they peer through the uncertainties to discern the slightest indications of where the "truth" lies and decide accordingly. In doing so they operate by a principle similar to that which guides the golfer, a principle that depends for its reasonableness on a similar value-laden calculation. Their "scientific bias" makes them willing to take risks when their "facts-only" calculation indicates little or no risk, an approach that may seem reasonable to them by the calculation that the prospect of gain, in the event they are proved right, is sufficient to warrant risking the loss, in the event they are proved wrong. In the context of pesticide regulation, calculations of this sort generally translate into according more weight to the prospect of economic benefits, obtained by following the "the factual indications only" and ignoring the uncertain indications of risks to health, than to the prospect of the deleterious health effects that will occur in the event those same indications prove misleading.

In short, risk assessors who strive in the face of mild uncertainty to be strictly neutral and to attend only to the facts, cannot attain the neutrality to which they aspire. Whether they recognize it or not, any decision they reach will depend for its reasonableness on the relative worth of the various values that their ostensibly value-free approach puts at risk.

We have said that value frameworks influence risk estimations by "framing" the assessment process. A result is that assessors with different value frameworks may frame the assessment process differently and be led by this to markedly different estimations of risk. A critical aspect of an assessor's way of framing the assessment process is his or her adoption of "argumentative strategies." Three such strategies, which were alluded to above, concern the assessor's working assumption regarding the degree of scientific rigour it is appropriate to employ and to expect from others, the assessor's assignment of the burden of proof, and the assessor's attitude toward risk-taking. These are described in more detail below.

Scientific Criteria of Causality

In a risk assessment context, each scientific issue lacking statistically reliable data requires the making of a judgement which ostensibly relates to the scientific data, but which also registers in the risk estimation by raising or lowering the calculated risk. This judgement thus serves a dual function: first, it weighs one body of evidence against another, deciding which to take as determinative; second, it constitutes a decision whether to guard against under-estimating or over-estimating the risk. In this way, what appears to be a purely scientific decision also becomes a public policy decision.

To illustrate, one influential issue in the risk assessment of alachlor was the relative importance of adenoma (benign glandular tumour) formation in adenocarcinoma (cancer) induction (*Report*, pp. 63, 64). It was uncertain whether the benign tumours found in the rat tissue were the precursors of carcinogenic tumours. While some evidence was available that they were not, a decision to this effect would lead to the exclusion of benign tumours from statistical analysis of the data on carcinogenicity and thus to a lowering of alachlor's calculated risk. Whether officially recognized by the analyst or not, this decision would express a resolve to guard against risk over-estimation rather than under-estimation since it would be a decision to exclude potential risk indicators from the calculation.

The decision whether to assume a causal relationship might have been seen to rest entirely upon the adequacy of the available information, for example, whether it satisfies the criteria for statistical significance. In rigorous, publishable science, the need for the kind of judgement illustrated in the preceding paragraph is often obviated by the application of the standard criterion of causality (95% confidence). This ensures that the correlation in question is statistically significant, i.e., that a *causal* relationship might be reliably inferred. Invocation of a rigorous criterion of causality, coupled with other stipulations designed to guarantee a completely solid evidential basis for conclusions drawn, yields what is sometimes referred to as "conservative science." In the Hearings, restricting oneself to conservative science was occasionally referred to as "scientific prudence." A less rigorous criterion, for example, "weight of evidence," yielding what is sometimes referred to as "liberal science," will obviously result in conclusions that are less compelling.

The choice of criterion has a significant effect on the calculated risk and is thus not only a scientific but a social policy decision. If one insists on statistically compelling evidence before acknowledging the existence of a relationship which would signify serious risks, then in the absence of that evidence one must act on the alternate assumption, viz., that the relationship and its attendant risks do not exist. In this way, the decision to use a rigorous criterion of causality can be (even if unwittingly) a decision to assume certain risks. To the risk-aversive critic this looks like an implicit bias toward risk-taking. On the other hand, analysts such as those associated with HPB, who consciously adopt a risk-aversive policy, will be satisfied with less evidence supporting hazard and thus will use the less rigorous criterion of causality. To critics on the other side, such as Monsanto, this looks like an unreasonable fear of risk.

Both criteria were defended at the Hearings on the basis of the values paramount to each party. A particular value framework thus lay beneath each choice of criterion of causality, a choice which in turn directed judgements on statistically uncertain data. The dependence of this choice on the assessor's values is underscored in the alachlor debate by the fact that, for reasons to be explained below, the three major parties switched their respective choices of causality criterion when interpreting the data on metolachlor. HPB employed a "liberal" or less rigorous criterion to assess alachlor, and a "conservative" or rigorous criterion to assess metolachlor. Monsanto and, less obviously, the Review Board used the rigorous criterion to assess alachlor, and the less rigorous criterion to assess metolachlor.

Burden of Proof

The shift between a rigorous and a less-than-rigorous approach is controlled by, and reinforces, the assessor's second argumentative strategy: assignment of the burden of proof. In risk assessments, the ever-present uncertainties create a need to decide which side bears the burden of proof. The idea of burden of proof is especially familiar in legal contexts. In a criminal trial the burden of proof is placed upon the party seeking a guilty verdict rather than the party seeking to establish innocence. This is because a fundamental principle of Anglo-Saxon criminal law is that the uncertainties implicit in the process of proving guilt and innocence should favour innocence. The prosecution must prove guilt "beyond all reasonable doubt," while the defence need only raise the doubt. This methodological requirement expresses a prior value judgement that "it is better that many guilty persons go unpunished than that one innocent person be punished unjustly."

In a risk assessment debate, reference to a "side" bearing the burden of proof assumes that the assessment will concern a technology (in the present case, alachlor) which is defended by a proponent (Monsanto) as being either *safe* or at least sufficiently beneficial to warrant accepting the risks it presents. The question then arises, whether the proponent has a burden of proof to establish that the technology is safe (or safe enough), on one hand, or, on the other, whether the "opponent" (here, the government) bears the burden of proof to establish that the technology is unsafe. In a technology debate the assignment to one side of the burden of proof is in effect an assignment of advantage to the *other* side in the following respects: first, the latter is given the power to invoke the rigorous criterion of causality associated with conservative science, thus lending the air of scientific objectivity and care to its own position, and discrediting the claims of the opposing party, which is unable under conditions of uncertainty to put forward a completely convincing, scientifically compelling case. Secondly, the party without the burden of proof is able to retreat easily to a less-than-rigorous, liberal science in establishing the feasibility of its own position, and to do so without loss of credibility, since it is being no less "scientific" than the opposing party. In this way the invocation of rigorous science in risk estimations, coupled with the *policy* decision concerning the burden of proof, permits *one* side in the debate (but not the other) to give what appears to be a definitive and reliable estimation of risk— even in a context where the uncertainties in the data are so great that, if a burden of proof were not assigned, a responsible scientist would have to withhold any judgement.

Thus, the debate between those who believe that the burden of proof should rest with the side claiming that the product is safe, and those who believe it should rest with the side claiming that the product poses unacceptably high risks, has implications for the issue of scientific standard of proof and for the resulting estimate of risk. Because a decision concerning the burden of proof sets the rules for the risk calculation in such a way as to make it much more difficult for one of the parties to prove its case, it in fact becomes a policy decision to accept or reject the risk itself. The underlying policy issue, analogous to that of unjust conviction, is put by the following question: Which is better? That many unsafe products go unrestricted, in order to ensure that no safe products be banned? Or that many safe products be restricted, if that is the price of ensuring that no unsafe products are on the market?

This regulatory policy choice was a primary factor behind the conflicting positions advanced in the debate concerning alachlor.[18] The government's concern for public safety prompted use of a less rigorous standard of causality, which led it to interpret the studies submitted by Monsanto as establishing that alachlor was carcinogenic in rats and mice. Since the animal cancers were produced at levels of exposure humans also might experience, the government concluded *in the absence of clear data to the contrary* that similar risks to humans should be assumed. On the other hand, Monsanto's concern for market security prompted use of a rigorous standard of causality, which led it to interpret the data as not warranting the conclusion that the levels of exposure expected from alachlor would cause cancers in humans. Since the studies did not establish that humans would be exposed to levels of alachlor similar to those that produced cancer in rats, it should be assumed *in the absence of clear data to the contrary* that alachlor would not cause cancers in human beings. The conclusions drawn by the two parties about the levels and the acceptability of risk were significantly influenced by their prior (and opposing) policy assumptions concerning the burden of proof, and these assumptions translated into controlling argumentative strategies in the risk debate.

The Review Board, in the development of its own case, was required to determine whether the government or Monsanto should shoulder the burden of proof. As we shall see in Chapter IV, in the end it accepted Monsanto's view that the government had the onus of proof and had failed to meet this onus.

Whether to be Risk-Aversive or a Risk-Taker

The third component of the frame by which risk estimate data are interpreted is the attitude toward risk-taking adopted by the assessor. The issue concerns where the assessor will stand on a spectrum marked at one end by being extremely risk-aversive and at the other end by being an unblinking risk-taker. In the alachlor debate the risk-takers had in view economic benefits and the risk-aversive assessors tended to give priority to avoidance of threats to human health. Monsanto's stance was that of the risk-taker; HPB was risk-averse. The Review Board appeared to vacillate but in the end came down decisively on the side of Monsanto. By staking out a position concerning risk-taking, each party to the debate armed itself with a distinctive argumentative strategy that guided its interpretation of the evidence before it and its response to claims advanced by the other parties.

This distinction between risk-taking and risk-aversive attitudes is generally not recognized in risk management contexts dominated by risk-benefit approaches. The reason for this is that the typical risk-benefit formulae reduce risks and benefits to one common denominator of value. So, for example, Monsanto clearly did not see itself as adopting a risk-taking attitude. Rather, it saw itself as choosing the lesser of two *risks*—the possible risks to human health and the evident "risks" to the agricultural economy. From this point of view, all utility maximizing choices are "risk-aversive." Thus, as we explain in greater detail in Chapters V and VI, the distinction between risk-aversive and risk-taking attitudes is closely related to scepticism with regard to the instrumental conception of rationality. Here, however, we follow the conventional use of these terms, so that, where health and economic benefits are the opposed values, to priorize the former is to be risk-aversive; to priorize the latter, a risk-taker. This makes sense in a regulatory context because the need to protect health is the *raison d'être* of the regulations.

It may seem that assessors' positions regarding risk will be dictated by their positions regarding the two previously mentioned issues, scientific standard and burden of proof, since, as discussion of these latter issues has indicated, to impose a burden of proof on the side which priorizes human health *is* to adopt a risk-taking stance. However, it is conceptually possible for an assessor's stance regarding risk to be an independent variable which controls the position taken regarding burden of proof. This matter is discussed at some length in Chapter VI.

The Impact of the Assessor's Value Framework on the "Frame"

The various elements of our alternative model of risk assessment and the significant relations among them have now been described. We have developed an account of the way risk estimations operate with an interpretive framework which takes up the data science has to offer, and we have indicated how the shape of that framework reflects the assessor's value framework. To this point, however, our account has focused on the elements of the alternative model, statically considered. We want now to represent the model dynamically, by exhibiting the interplay of the various elements as they guide the assessment process toward its final conclusion, along with the interplay between opposing estimates in a technology debate, each of which is founded on a distinctive value framework. We shall first exhibit the dynamics of the model by indicating how the various components interacted in the alachlor debate, and shall then generalize from this to offer a formal account of the dynamic model.

HPB officials were well aware of the connection between scientific criterion of causality, placement of the burden of proof, attitude toward risk, and public policy. In their minds, the "cause for concern" which the data gave was a sufficient scientific basis for cancelling alachlor's registration precisely because the onus was on Monsanto to produce the more rigorous case—and the company could not. In a few cases, feeding studies supported a conclusion that alachlor was responsible for carcinogenic tumour development, but the evidence, although biologically significant, was not statistically significant. HPB consistently treated the studies as suggesting that alachlor poses a health risk, despite the lack of statistical significance. HPB officials were attempting to implement a risk-aversive policy with respect to pesticide-related cancer: for them, the value of a safe environment outweighed the benefits secured by putting that environment at risk.

That HPB's regulatory science was not neutral was clearly articulated during the Hearings. For HPB, the regulatory context required "a bias in terms of [health-related] conservatism." The principle used, according to Dr. Ritter, was that originally put forward by Dr. Doull, whom Ritter called the "current father of toxicology": "Where science cannot provide the answer, prudence must" (*Hearings*, p. 3070). Ritter claimed that HPB exercised "toxicological prudence" when interpreting the rat and mouse alachlor studies and that it was up to the Minister of Agriculture to exercise a form of "regulatory prudence" (*Hearings*, p. 3071). Relevant to the government's concern was also the carcinogenic nature of the risk (*Hearings*, p. 3509). The government's conclusion—that the submitted data clearly showed alachlor to be a potent animal carcinogen even at relatively low levels—was thus based upon interpretive assumptions chosen specifically for their prudent handling of uncertainty, that is, for their capacity to implement a policy of minimizing environmentally-induced cancer in rural communities.

However, HPB officials appeared to adopt the opposite standard when they interpreted the risks of metolachlor. While metolachlor was regarded by HPB analysts as safer than alachlor, they appeared to enhance the difference between the two by not applying the principle of prudence to the interpretation of metolachlor data. Thus, while alachlor was found to be an animal carcinogen and a potential human carcinogen at exposure levels equal to those experienced by farm applicators, metolachlor was found to be neither an animal nor, by extrapolation, a human carcinogen. In this way, it seems that HPB officials were able to bypass the difficulty of quantifying their comparative risks. They allowed to remain on the market the product they felt, on the basis of a qualitative reading of the data, to be the safer (*Hearings*, pp. 3308, 3311).

Like the Government, Monsanto made its values and the related issues, standard of scientific proof, burden of proof, and attitude toward risk, clear at the outset. "The real issue," Monsanto argued, "is whether regulatory decisions should be based on overly conservative assumptions of questionable scientific validity. This unnecessarily puts the Canadian farmer at a competitive disadvantage in the world markets and impairs the Canadian economy by preventing the Canadian farmer from utilizing needed agricultural chemicals" (*Report*, p. 13). The company argued for a reading of the data that minimized health risks in an attempt to minimize the loss of benefits that might result from risk overestimation. It thus geared its assumptions to guard against over-estimating rather than under-

estimating the risk by placing the burden of proof on the government to prove safety, arguing that this proof should be scientifically
rigorous, and by using a "typical-case" rather than a "worst-case"
scenario to predict operator exposure. Its framework of value was
thus clearly and consistently market-related.

However, like the Government, Monsanto switched to a different
set of assumptions when it assessed the data on metolachlor and
four other chloracetanilide compounds, which the company produced in confidence at the Hearings. For metolachlor and two of the
compounds that had produced tumours similar to those in the
alachlor studies, Monsanto argued that the data clearly showed carcinogenicity. This argument could only be made by upgrading the
risk associated with uncertainty of chemical action—an apparent
strategic attempt to make the strongest possible case for a finding
that its own product was no riskier than the competing product that
the government was not ready to withdraw from the market. This
finding would effectively eliminate the management option of banning alachlor while leaving metolachlor on the market. So, with
respect to metolachlor, Monsanto was willing to draw the very same
kind of statistically unsupported (but biologically defensible) conclusions it had ruled out with respect to its own product.

The Review Board was equally inconsistent in its reading of the
data on the two chemicals, but was less clear about the values its various readings upheld. On one hand, the Board appears to have interpreted its activities in the terms of classical risk assessment, as neutral and scientific. Thus, in the *Report* (p. 27), the Board described the
Hearings as primarily scientific, and the choice of scientists as Board
members appropriate in that context. "Better science" directed its
decision to use the "weight of evidence" method of risk assessment
over the quantitative modelling urged by Monsanto.[19] And the scientific context seems to have been assumed to be the only source of
criteria for "reasonableness" in risk assessment judgements, an
assumption of a piece with the Board's inability to see the need for a
clear accounting of the other criteria related to the regulatory context
that it used in making judgements of its own.

On the other hand, as just implied, in its estimate of alachlor's
risks the Board made many judgements during the course of the risk
assessment which drew values from contexts other than the scientific
and which were never explicitly defended. We have inferred these
from the general trend of concerns and opinions brought forward
during the Hearings and from the Board's defence of judgements
made throughout their *Report*. The Board expressed, for example, a
consistent concern with the "unjustified costs and ... unjustified
results" entailed by HPB's "unreasonable worst-case assumptions."

The Board felt that HPB's unreasonable assumptions had led the Branch to overestimate alachlor's risks. "Regulatory prudence" ought to dictate "the use of *reasonable* worst-case assumptions ... when science cannot provide precise answers" (*Report*, p. 62).

Regulatory prudence was clearly not identical with scientific prudence. During the Hearings, Dr. Rowe, with Dr. Sielkin, Dr. Freshwater and others, argued that regulatory and scientific prudence were sometimes at odds. Substantial costs would accrue if the safety estimates were unreasonably "prudent," that is, if the margin of safety added to each of the factors in the risk assessment was too high. The margins built into the government's risk assessment to cover the uncertainties in the studies and to protect public health, had costs "in the regulatory sense in terms of opportunities foregone" (*Hearings*, p. 3476). In other words, the loss of certain benefits through unreasonable widening of the margin of safety was irrational. Dr. Rowe (Review Board) thus inquired of Dr. Ritter (HPB) if the latter's exercise of toxicological prudence may not have *prevented* the regulator (Minister of Agriculture) from exercising regulatory prudence (*Hearings*, p. 3072). The inference is that in this situation upgrading risks can be as or more harmful than downgrading them, in light of the unforeseen consequences of the decision and the benefits at stake. For the Board, regulatory prudence thus appears to have required cognizance of the benefits involved in the adjustment of the margin(s) of safety.

That the Board made this connection is supported by its use of the term "reasonable" in the concept of a "reasonable worst-case exposure scenario." The criterion of "reasonableness" here is not that of scientifically reliable prediction or of "due regard" for human health, but rather of socially acceptable regulation. As we show in some detail in our review of the Board's case below, for it a reasonable worst-case scenario is one that, above all, is fair to the regulated. As such, it is very close to Monsanto's "typical case." It assumes responsible rather than actual use of the product, on the grounds that if farm applicators do not follow label instructions, do not wear full protective clothing, or allow spills and accidents to occur, etc., it is not of regulatory concern.

Contrary to the classical model and to the Board's view of its own position, then, its estimate of alachlor's risk, like Monsanto's, was charged with distinctively market-related values. In holding that in regulatory situations maximum fairness to the regulated and the costs of potential over-regulation ought to weigh most heavily in the determination of risk, the Board adopted Monsanto's priorities. As we show, this led the Board to interpret uncertainty in the data so that the values thus priorized would be least at risk. It thus elevated

the social value of maintaining the current chemically-based agricultural economy over that of reducing the incidence of cancer in the
rural community. At least it ensured, perhaps unwittingly, that the
interest in continuing with chemically-based agriculture would be
accorded significant weight in the management of alachlor's risks.
While this may or may not be the right priority, its cover of scientific
neutrality tends to insulate the issue from the forum of public
debate.

What complicates the Board's overall position, like that of the
Crown and of Monsanto, is the way the Board used these values
quite differently in its interpretation of the data on metolachlor. For
metolachlor, the Board employed a context of human health to argue
for a reading that upgraded the risk assessment (*Report*, p. 70) and
thus minimized the difference between the two products. In this
way, the Board eliminated the need to compare the risks of alachlor
and metolachlor, which it clearly felt disinclined to do. This comparison, all parties agreed, was impossible to quantify due to the differences in test conditions and the uncertainties in extrapolating animal
to human data. However the Board held the stronger position that
such a comparison was simply impossible, and this also ruled out a
qualitative assessment.[20]

The foregoing is an account of the differing ways the three parties
to the alachlor debate framed the assessment project to reach differing conclusions based on the same data base, and of how these
frameworks reflected the differing values held by the parties to the
debate. Based on these remarks, we offer here a formal account of
the dynamic relations among the framing components. This will
require clarifying how the three components interrelate to form a
distinctive way of framing the risk assessment process—how the
assessor's value framework dictates the frame and how the frame
guides interpretation of the data.

Our model defines the risk assessment process in its most principled form. We say this because the process can of course be subverted and the data made to support independently determined
ends. This is a tactic that we want to distinguish clearly from the
interpretation of scientific data in light of a framework of values. In
the latter process, publicly defensible values are selected to design
scientifically credible tests and to interpret uncertainties in the resul-
tant data, leaving the conclusion pending full assessment of the data;
in the former process, these techniques are manipulated so that the
scientific information is made to support conclusions that are prede-
termined and may be unrelated to the data.

For our model we shall retain the simplifying image of a technology proponent, whose primary value is economic benefit, and an

opponent, whose primary value is human health. Then, as suggested above, the proponent will adopt a risk-taking stance toward the technology: there will be a bias toward accepting the health risks, of course, but, more importantly, toward deciding "close calls" in the risk assessment in ways that minimize the calculated risk. This risk minimization strategy will be facilitated by assigning a burden of proof on any particular issue to the side which maintains a position emphasizing the risks, directly or in its final effect. Assignment of the burden of proof to that side is made manifest in the invocation of a rigorous criterion of causality, and more generally in application of the standards of conservative science to the position taken by that side. The effect of these three argumentative strategies, all dictated by the priorizing of economic benefits, is to produce an estimate that minimizes the calculated risk of the technology and thus protects the priorized value.

By contrast, the opponent will adopt a risk-aversive stance toward the technology: there will be a bias toward emphasizing the health risks and toward deciding close calls in the risk assessment in ways that emphasize the calculated risk. The argumentative strategies for accomplishing this will be:

1. Assignment of a burden of proof on any particular issue to the side which maintains a position minimizing the risk, directly or in its final effect.
2. Application of the high standards of conservative science to the position taken by that side.
3. Adoption of a risk-aversive stance when uncertainties cloud issues that need to be resolved before an estimate of risk can be made.

By framing the risk assessment process in this way, an estimate will be reached that emphasizes the risk and thus protects the priorized value, human health.

We have, of course, caricatured the framing process to exhibit its power. Nevertheless, the caricature strikingly reflects many aspects of the actual alachlor debate. While the government, Monsanto, and the Review Board may not have all been equally self-conscious of the way they worked within the frame, all used the components of the frame to protect their own values, or to ensure that the final result would be consistent with their value framework. Given the uncertainties, and the presence of inherently normative issues in the risk assessment, it could not have been otherwise.

A final comment regarding the relation of the classical model of risk assessment to the alternative model sketched out in this chapter. The classical model stresses the indispensability of basing risk estimations on reliable scientific data. We have no quarrel with that. To repeat a point made earlier, bad science results in bad risk assessments. The point of difference concerns the *sufficiency* of science for risk assessments. What the alternative model adds is the indispensability of an interpretive point of view which reflects the assessor's value framework and which frames the assessor's attempt to determine what the data imply concerning a product's risks. This difference is momentous. It implies that risk assessment cannot be, as the classical model maintains, value-free. It also implies, as we shall argue in Chapter IV, that "risk" is not an objective property of the product which is said to pose the risk.

III

The Arguments of the Government and Monsanto

The Government's Case against the Acceptability of Alachlor's Risks

The Carcinogenicity of Alachlor

The government's decision to cancel alachlor's registration was entirely consistent with the chosen standard of safety with which it appeared to have interpreted the results of the animal test and applicator exposure data submitted to it by Monsanto. This standard, never officially adopted by the government but appealed to at several points in the Hearings, was to eliminate or at least reduce to a minimum the risks of cancer. HPB's interpretation of the data was significantly different from that of Monsanto and the Review Board. By 1983, the data available to the Branch, taken from five separate tests on rats (four carried out as a series in the same laboratory), showed alachlor to have caused statistically significant cancers (carcinomas) at the highest dose (126 mg/kg/day) at each of two primary sites, the nasal turbinates and the stomach. Lower doses also induced tumours, two cancers at 42 mg/kg/day in the nasal turbinates and one stomach tumour at 2.5 mg/kg/day. A large number of noncancerous lymphatic tumours (adenomas) were induced as well in the nasal turbinates in all but the two lowest doses (0.5 and 2.5, where there were none and one, respectively). Noncancerous brain tumours were found at 42 and 126 mg/kg/day; these were later determined to have originated in the nasal turbinates.

The alachlor test results were tabulated by the Review Board as follows:

Laboratory Test Protocol

BD-77-421   Doses were 14, 42 and 126 mg/kg/day for 93, 86 and 90 test animals respectively for 25 months. (For the first year a stabilizer was used which was found to be carcinogenic and was then switched.)

ML-80-224-I   Dose for 86 test animals was 126 mg/kg/day for 25 months.

ML-80-224-II   Dose for an unstated number of test animals was 126 mg/kg/day for 5-6 months. (Only ocular lesions examined.)

ML-80-224-III   Dose for 63 test animals was 126 mg/kg/day just for 6 months; the animals were then fed an alachlor-free diet until 25 months had elapsed.

ML-80-186   Doses were 0.0, 0.5, 2.5 and 15 mg/kg/day for [number of control animals unstated], 92, 92 and 93 test animals for 25 months.

Results

For the precise relationship between dose and adenomas or carcinomas, see *Report*, Figures 2 and 4, pp. 54, 57.

1) Incidence of the relatively rare nasal adenomas and adenocarcinomas:
   - Considering only adenocarcinomas, statistically significant cancers developed at the highest dose level, 126 mg/kg/day, only in males (ML-80-224-I).
   - Considering the combined incidence of the two tumour types, statistically significant tumours developed at the 126 mg/kg/day dose level in both males and females (ML-80-224-I and ML-80-224-III), and at 15 mg/kg/day in both males and females (ML-80-186).
   - At the lowest dose, 2.5 mg/kg/day, equal to the highest human exposure level projected by HPB, one nasal adenoma developed (ML-80-186). Although not statistically significant, this tumour was considered to be part of the statistical dose-response relationship and thus biologically significant.

2) Incidence of rare stomach carcinomas:
   - At 126 mg/kg/day in males and females (BD-77-421), and just in females (ML-80-224-I), statistically significant stomach cancers were found.
   - At 42 mg/kg/day, one female developed a stomach cancer (BD-77-421).

- At the lowest dose level, 2.5 mg/kg/day, one male developed a stomach cancer (ML-80-224-I).
- While these latter two cancers were not statistically significant, their rarity was believed by HPB to give them biological significance.

3) The presence of neoplasms in the brain extended the range of affected tissues to three. Subsequent analysis showed the origin of these neoplasms to be the nasal turbinates; however the metastases of the nasal adenomas were considered as well by HPB to be signs of the potency of alachlor's carcinogenic activity.
   - At 126 mg/kg/day, these neoplasms were found in 2 males (one each in experiments BD-77-421 and ML-80-224-I); and in 1 female (ML-80-224-I). At 42 mg/kg/day, they were found in 2 females and 2 males (BD-77-421).

An 18-month mouse study also showed a statistically significant increase in lung tumours only in the females at the highest dose of 260 mg/kg/day. A difficulty with the assessment of statistical significance, however, was that it depended upon use of a concurrent control group which had a lower incidence of lung tumours than the historical norm.

HPB analysts described their approach to data analysis as the protection of human health in the Canadian product regulation process. This process supplies analysts with data generated under a variety of test conditions, on the basis of which they must establish the absolute and relative potencies of chemicals under federal consideration. That the data usually offer a far from ideal basis for comparison does not reduce the need and pressure for comparison. Branch analysts therefore make a practice of reviewing the data as a whole, whether it has been produced under the same or different laboratory conditions, and drawing general "weight of evidence" conclusions on its basis. In doing this, they place a high importance on the presence of dose-response relationships, which show up more clearly as the frequency of tumour incidence and the size of the test population increase. With very rare tumours and small test populations, as in the case of the rat stomach tumours above, a classical dose-response curve is unlikely to be found: "as the rarity of the tumour increases, as the likelihood of seeing the event decreases, then in fact if the group size is held constant . . . failure to see the event does not necessarily suggest that the event is not taking place" (Dr. Ritter, *Hearings*, p. 3065). Analysts also assess the full range of quantitative and qualitative data characteristics:

the kind of tumour which was induced, the frequency with which that tumour was induced, the relative rarity of the tumour, the frequency with which it would occur in historical control populations, and whether or not there is any evidence of metastases, whether or not there is any evidence of early induction of tumours with treatment [i.e., exposure to the chemical under review], the number of different studies in which the same kind of effect or similar kind of effect may appear, the opportunity for exposure, the probability that exposure reduction measures can be applied in any kind of a meaningful context. (Dr. Ritter, *Hearings*, p. 3148)

HPB analysts found the alachlor data to exhibit a dose-response relationship at two sites, the nasal turbinates and, most importantly, the stomach. The rare stomach tumours were found in high numbers at the highest dose level in separate experiments. Concern over these findings was augmented by the presence of tumours in the brain, the probable carcinogenic effect of alachlor on the mouse, and the presence of cancers in rat tissue at the low doses to which farm applicators would be exposed. As we discuss below, the absence of a margin of safety was seen to be particularly important in light of the various forms and degrees of uncertainty in the case. Coupled with the assumption that metolachlor was a safe alternative, this evidence led the Branch to determine that alachlor was in fact a potent carcinogen even at relatively low levels in the rat and possibly a carcinogen in the mouse,[21] and to recommend the product's cancellation. In concluding that there was reason to assume that alachlor was a "probable human carcinogen," HPB officials had adopted the rating scale of the International Agency for Research on Cancer (IARC), which classifies substances as "known, probable, possible, or non-carcinogens" in humans. The IARC also classified alachlor as a "probable carcinogen."

The Carcinogenicity of Metolachlor

A confusing and controversial issue in the decision to cancel was whether the government considered the risks of alachlor to be unacceptable by the terms of some "absolute" standard of safety, or unacceptable only relative to the risks of alternative weed-control methods. In this case the major alternative control product in use by farmers, and alachlor's major competitor, was the chemically closely related product, metolachlor. It was clear that alachlor's cancellation would lead to increased use of metolachlor as a replacement herbicide, so the natural question, raised both by Monsanto and the government, was whether it posed any lower risk than alachlor.

HPB officials were concerned about the safety of metolachlor and asked that rat studies on its toxicity be reviewed by toxicologists within the Branch. Data on metolachlor available to the Branch included an IBT mouse study, a 1979 IBT rat study (determined by the EPA to be invalid as a chronic study due to mistakes in the dose levels, but to contain some data useful to the determination of metolachlor's carcinogenicity), and a 1983 Hazelton study (*Report*, p. 69, *Hearings*, pp. 4007-09).

The critical diagnostic study was the Hazelton study, carried out in two parts. In the first, 69 male and 69 female rats were fed weight-adjusted doses of metolachlor which averaged overall to about 150 mg/kg/day (*Hearings*, p. 3920). This study produced two males with nasal carcinomas and a third with a nasal adenoma. In the second, rats were fed increasing doses of metolachlor, 0, 30, 300, and 3000 ppm (0, 1.5, 15, and 150 mg/kg/day) respectively for 24 months (*Report*, p. 69). Hepatocellular carcinomas and liver nodules were found in the following ratios: 0/50, 1/60, 2/59 and 7/60 in females and 2/60, 2/57, 2/60 and 9/60 in males.

In the IBT study, metolachlor induced hepatocellular carcinomas and "hypertrophic-hyperplastic nodules" in the female liver in the following ratios: [no data given in the *Report* for the control] 1/54, 1/58, 3/60 and 11/60 at [0], 30, 300, 1000 and 3000 ppm ([0], 1.5, 15, 50 and 150 mg/kg/day) of metolachlor, respectively. This study also produced a small number of very rare cancers, angiosarcomas, one in 1 male at 15 mg/kg/day and one in 1 female at 150 mg/kg/day (*Report*, p. 69).

During 1983 and 1984, HPB carried out a review of these data and determined on the basis of the "preliminary evidence" that neither the cancers in the nasal turbinates and the liver nor the angiosarcomas could be attributed to metolachlor treatment. The Branch concluded that the chemical was not an animal carcinogen and informed Agriculture Canada that "repeat chronic studies on metolachlor [have] not demonstrated that this product has carcinogenic activity under the test conditions of these studies even though it is structurally similar to alachlor" (*Report*, p. 39). Analysts at the Hearings explicitly rejected the quantitative risk assessment of metolachlor submitted by Monsanto, which had concluded that the product was an animal carcinogen.

The risk aspect of HPB's decisions on the two chemicals can be seen by reviewing the Branch's approach to uncertainty. In each data set, and we suspect in general for regulatory science, few issues were evidentially secure. Many extra-scientific judgements, some of them key to decisions on presence and degree of carcinogenic activity, were made necessary by the high degree of uncertainty in the data. This much seems to have been agreed upon by all parties. As we have discussed, HPB analysts were well aware that their health protection mandate produced a risk aversive approach to uncertainty, and throughout the Hearings they defended the Branch's conservatism as policy-based. Their use of health protectiveness as an analytic tool was frequently challenged by Review Board members, however, as violating the parallel mandate, fairness in product regulation (*Report*, p. 62).

We list below the most important of the judgements made by HPB concerning alachlor and metolachlor:

HPB-A1. Alachlor: The nasal turbinate tumour was an "extremely uncommon tumour in untreated control groups" (Dr. Ritter, *Hearings*, p. 3111). The stomach carcinomas "occur infrequently at best, some would argue very rarely" (Dr. Ritter, *Hearings*, p. 3066).

HPB-M1. Metolachlor: The Branch accepted the judgement of an outside expert that the nasal turbinate tumours were not significant. (Since 1983, these tumours had been looked for more frequently; however, this procedural change had not made a difference to the Branch's own findings. The Branch therefore sought the opinion of an outside expert on the question of the significance of the nasal tumours: two cancers and an adenoma at the highest dose.) "We had some problems in deciding whether that [the three nasal turbinate tumours] was significant or not. . . . What we did therefore was to go to the person who was reputed to be the expert" (Dr. Clegg, *Hearings*, p. 3928).

HPB-A2. Alachlor: According to the Review Board, HPB felt that the data showed a dose-response relationship at multiple sites in both sexes (*Report*, p. 40). It is clear that the stomach tumours caused HPB analysts the most concern, due to their unquestionable rarity and the number that occurred at the highest dose level (*Hearings*, p. 3971-73). Given their rarity, single instances at lower dose levels were interpreted as indicative of a dose-response relationship. It is not as clear just what significance HPB accorded to the nasal tumours. The Review Board *Report* indicates that the Branch considered both adenomas and adenocarcinomas (*Report*, p. 55). Further, the Branch could not have interpreted a dose-response relationship to hold for tumours at the nasal turbinate site without inclusion of the adenomas. That HPB did in fact infer a carcinogenic response from these tumours is shown by the Branch's response to the six-month exposure studies, which it saw as "substantial evidence of an oncogenic effect" (Letter of June 25, 1984 from Dr. Morrison to Dr. McGowan, *Report*, p. 37). These studies produced 10 adenomas in 17 test males, with no cancers, and 19 adenomas in 46 test females with one cancer. It is the adenomas rather than the adenocarcinomas which appear in significant numbers in this study.

HPB-M2. Metolachlor: The Branch interpreted neither the nasal tumours nor the liver tumours as exhibiting a dose-response relationship. The nasal tumours occurred only at the highest dose level. The pattern of liver tumours, however, was less clear. HPB's decision *not* to combine the liver carcinomas with the liver nodules resulted in a tumour incidence which analysts considered either to be counter to a dose-response (2, 1, 3, 2 in males at increasing dose levels) or to be slightly ambiguous but overall not far off background levels (0, 0, 0, 2 in females) (Dr. Clegg, *Hearings*, pp. 3924-25).

HPB-A3. Alachlor: It can be argued that although firm proof was not available, the nasal adenomas were considered as precursors of carcinomas and combined for assessment in the analysis of alachlor data. Reinforcement for this position can be found in the Hearings testimony, although the issue is clouded by arguments as to the *necessity* of adenomas for the generation of adenocarcinomas—arguments that become salient in the interpretation of metolachlor data, where an equivalent number of cancers were found in the absence of the large number of adenomas.[22]

HPB-M3. Metolachlor: Although expert opinion was divided on whether liver nodules are precursors of liver cancers, HPB chose to treat them as nonprecursors. Here, HPB did not combine benign and carcinogenic tumours for assessment (*Hearings*, p. 3925).

HPB-A4. Alachlor: The statistically non-significant stomach carcinomas at low doses were, in the absence of evidence to the contrary, considered part of the dose-response pattern and hence, biologically significant.[23]

HPB-M4. Metolachlor: In light of the absence of a dose-response relationship at the nasal turbinate site, it was decided that the two nasal cancers ought not to be considered biologically significant (*Hearings*, p. 3928; *Report*, p. 60).

HPB-A5. Alachlor: The discovery that the four brain tumours (ML-80-224) were in fact metastasized nasal turbinate carcinomas was considered as further evidence of the carcinogenic potential of alachlor—"in medicine, one of the hallmarks of cancer is metastasis" (Dr. Ritter, personal communication, October 23, 1989)—rather than as diminished evidence in light of its reduction of cancer sites from three to two.

HPB-M5. Metolachlor: In light of the absence of a dose-response relationship, HPB analysts initially did not consider the angiosarcomas or the range of tissues affected to weigh in favour of metolachlor's carcinogenicity (*Hearings*, pp. 3924-25; *Report*, p. 60).

HPB-A6. Alachlor: The range of affected tissues, the sheer number of tumours, and the low dose which produced a stomach tumour were seen as evidence adding to the weight of the dose-response relationship in demonstrating alachlor's carcinogenicity.

HPB-A7. Alachlor: HPB's interpretation of the mouse data was based upon the assumption that while the statistical significance of the data was questionable, "this study cannot be regarded as negative and thus provides supportive evidence for the positive results noted in the rat studies" (*Hearings*, p. 3073). This conclusion was shared by the EPA (*Hearings*, p. 2931). Further, HPB argued, concurrent controls reflected the effects of the same laboratory conditions as the test animals experienced, and therefore provided a better basis than historic controls for calculations of statistical significance.

HPB-A8. Alachlor and Metolachlor: The Branch's final assumption concerned the way the dose-response relationship generated by these data should be extrapolated to human populations, in the absence of human data. HPB took the position that for regulatory purposes a carcinogenic response to a chemical in laboratory animals should be taken as determinative of the question of risk to humans. This assumption was disputed by Monsanto and contested with metabolism and epidemiological studies submitted at the Review Board Hearings. Neither the government nor the Review Board, however, accepted the scientific validity of these studies.

In their analysis of the alachlor data, HPB analysts employed a risk aversive stance each time a scientifically uncertain issue forced a personal judgement. In this way, the Branch developed a strong biological case for alachlor's risks from a body of data which, derived from a relatively small test population, was statistically weak overall. Analysts' confidence in their view was strengthened by what they saw, given the size of the test population, as an unambiguous sign of carcinogenicity: the presence in independent tests of large numbers of rare stomach cancers at the highest dose level. That the metolachlor data offered no such sign may explain, if not completely justify, HPB's far less conservative conclusion concerning that chemical.

To a large extent, analysts based their overall risk assessment of both chemicals on the presence or absence of a dose-response relationship. In neither data base, however, was that relationship unambiguous and, as the above analysis suggests, some of the judgements used to ascertain its presence in the alachlor data were contrary to those used to ascertain its absence in the metolachlor data. Assumptions M2 and M3, that liver nodules ought not to be considered as precursors to cancer, conflict with the parallel decision on alachlor nasal turbinate adenomas (A2 and A3) and could not be held up during cross-examination at the Hearings.[24]

The decision not to consider the liver nodules as carcinogenic had the effect of rejecting data which the Review Board later interpreted as establishing a dose-response relationship. As mentioned, this was a key decision since the Branch's approach to the metolachlor data as a whole was coloured by failure of those data to support a dose-response relationship. The decision resulted in the Branch's underplaying the potential significance of the number of liver cancers in test males and minimizing the possibility of a dose-response relationship in test females. The decision not to include liver nodules in the assessment, a (health) risk taking decision, led to the further decision not to interpret the liver cancers as treatment related—another risk taking decision.

There is apparent inconsistency as well in HPB's refusal to see the two nasal carcinomas and one adenoma as cause for concern (M1). As assumption A1 shows, this refusal was not supported by the Branch's own experience; the opposite assumption was used to support the case for alachlor's carcinogenicity. The difficulty experienced by the Branch with this judgement is itself testimony to the degree of uncertainty involved in the issue. Perhaps the tumours ought to be considered significant; HPB could not say. In the face of this uncertainty, the Branch chose to accept the position which denied the tumours' possible significance, a position, again, of health risk-taking.

The appearance of inconsistency here dissipates somewhat when one takes account of HPB's overall strategy: since it did not deem the removal of both alachlor and metolachlor from the market as feasible, alachlor could be removed only if metolachlor was in fact a safer alternative. From this point of view its downplaying of the latter's risks was motivated by a risk aversive approach to alachlor.

The shifts made by HPB in its estimation of metolachlor's risks were reflected in the difficulty Branch witnesses had in defending their risk estimation under cross-examination during the Hearings.[25] This is especially true of their initial judgements that metolachlor "is not an animal carcinogen" and "has not demonstrated ... carcinogenic activity under the test conditions of these studies" (*Report*, p. 39). More supportable was HPB's weaker claim, that although there is reason for concern about metolachlor's carcinogenicity, the weight of the evidence, interpreted qualitatively, showed metolachlor to be less dangerous than alachlor (*Hearings*, pp. 3308, 3311).

Despite this risk assessment, HPB officials held throughout the Hearings that the question of metolachlor's relative safety was not a determining factor in the recommendation for cancellation, because they considered the risk of cancer posed by alachlor to be sufficient by itself to justify that recommendation. Agriculture Canada officials also claimed this as their view, but were more ready to admit at the Hearings that the assurance from HPB that metolachlor was less risky had entered into their decision. This ambivalence on the part of the government appears to have been a major issue in the Review Board's decision against the government's cancellation order. The Board's perception that the government's decision-making procedures were out of the reach of public accountability was in large part a response to the government's inability to hold firm on the question of metolachlor's safety and on the importance of this information to the final decision on alachlor. The Review Board accepted Monsanto's views that the safety of alternatives was a necessary consideration in alachlor's risk assessment and that available evidence did not support the government's conclusion that metolachlor was a safer product. And, interestingly, as the following analysis suggests, the assumptions used by the Board and Monsanto to dispute the government's conclusion about metolachlor were opposed both to those of the government and to those used in their own analyses of alachlor.

Estimation of Exposure: Determining the Margin of Safety

Having established the presence of a dose-response relationship between alachlor and cancer in rats, analysts were required to calculate the levels of alachlor to which human applicators would be exposed and to measure these against the rat dose-response data. The difference between the lowest level which induced a tumour in rats (2.5 mg/kg/day) and the exposure level calculated for farm applicators and the rural public was seen by all parties as the "margin of safety." As we discuss in Chapter IV, just what each party meant by this term is not entirely clear. "Safety" seems in most instances to have meant "zero increase in the incidence of cancer in humans," since nowhere was risk translated into actual increased cancers. With a zero-risk standard, assumption of a margin of safety carries with it the assumption of a threshold level of exposure below which alachlor causes no cancers. The rat data were inconclusive as to the existence of a threshold, since the failure of the lowest dose (0.5 mg/kg/day) to produce tumours may reflect the small size of the test population. The need to invoke the threshold assumption was felt only by Monsanto and the Review Board, who argued for the presence of a "reasonable margin of safety" between the tumour-causing dose in rats and their calculated human exposure levels.

HPB's calculation of farm applicator exposure was 2.7 mg/kg/day, virtually identical to the lowest dose which induced tumours in rats (2.5 mg/kg/day). To the Branch, this offered no margin of safety and officials recommended that alachlor's registration be cancelled. Monsanto calculated a range of 0.000031-0.0000009 and 0.000016 mg/kg/day for farmers and commercial applicators respectively, and argued that their product posed no risk at all of cancer. After a review of the assumptions leading to these opposing risk assessments, the Board calculated an exposure level of 0.001-0.0001 mg/kg/day, estimates approximately three to four orders of magnitude lower than the lowest dose which had induced tumours in the rat studies. The Board considered this "to be a reasonable margin of safety" (*Report*, p. 84). Again, the last two arguments rely upon a threshold mode of action for alachlor.

The range of estimations of farm applicator exposure calculated by the three parties was thus 2.7 to 0.0000009 mg/kg/day. The wide differences in analysts' exposure estimates resulted from their adopting differing positions on a limited number of key issues, none of which could be unambiguously resolved by referring to conclusive data. The first of these concerned the method for estimating the total body dose of alachlor which applicators would receive. Two techniques were used on selected farm applicators, neither of which measured the critical parameter, blood level. The first was the patch test, skin patches which collected ambient alachlor over a fixed period; the second was biomonitoring, urine samples taken from selected alachlor applicators. These were the only sources of exposure data obtained from human subjects. Further assumptions were made on the basis of anecdotal evidence and animal test data.

Both methods were employed to monitor the amount of pesticide to which applicators were exposed in field tests. In conducting these tests, decisions had to be made concerning the kind of protective clothing applicators would wear and other features of the pesticide application procedure. Exposure estimates were highly sensitive to these decisions, which in turn reflected differing assumptions about the purpose of such tests. HPB held that the tests ought to reflect actual conditions in the field, whereas Monsanto argued that the tests ought to measure exposure incurred only by those applicators who follow the instructions on the product label. The patch test method required a further assumption about the percentage of alachlor deposited on the skin which would be absorbed into the blood stream. And the biomonitoring method required an assumption about the metabolism of the product which would allow levels excreted in the applicator's urine to be correlated to blood stream levels.

The figure calculated by either method was then subject to a decision about amortization, that is, whether to average short-term exposure levels over an individual's lifetime, on the assumptions that (a) because the effects of the chemical in question are cumulative, cancer is related to total lifetime dose rather than shorter-term concentrated levels of exposure, and (b) a threshold of safety thus exists under which exposure to that chemical poses no risk of cancer. This proved to be a critical indicator of the value-orientation of the analysts, since almost none of the requisite information for a scientifically credible decision was available and it was ultimately the most important decision in the calculation of an exposure estimate.

It can be seen that these various assumptions required a rationale external to the context of scientific investigation, for two reasons. First, evidence provided by the witnesses at the Hearings showed the needed data to be unavailable and the available data often inconclusive. Second, as indicated above, some of the assumptions were not empirical in any case.

HPB made the following assumptions concerning exposure estimates (*Report*, p. 58):

HPB-E1. HPB used the only test data available to them at the time, patch test data.

HPB-E2. The exposed area of the applicator was assumed to be 75% of the body surface, which implied that no gloves or protective clothing would be worn. HPB reasoned that not all farmers could be assumed to wear protective clothing and that some gloves were known to fail.

HPB-E3. 100% of the product deposited on the skin was assumed to be absorbed into the blood stream. Although rhesus monkey dermal exposure studies available from Monsanto indicated that absorption levels ranged from 5.8 to 19.8%, HPB did not have confidence in their reliability. Dr. Malik, witness for Monsanto, confirmed under cross-examination that confidence should not have been placed in these studies (*Hearings*, p. 1387). In the absence of a scientific basis for transferring the ratio from monkeys to humans, the Branch argued for a "worst-case" assumption.

HPB-E4. The calculated daily exposure should not be amortized. The government rejected the claim made by Monsanto and several expert risk assessment witnesses at the Hearings that the animal studies supported a "threshold" level of carcinogenic response to alachlor below which there was no risk. Just because there was no *observed* effect below certain dose levels of alachlor, government analysts reasoned, it cannot be concluded that there is no effect in reality. They argued that the non-appearance of carcinogenic effects might be nothing more than a function of the small sample size or other variables, and they reached the alternative conclusion that the level of cancer risk is directly proportional to the level of exposure (*Hearings*, p. 5288). In their mind this was supported by the test results. The rats fed alachlor each day for 5-6 months and then a normal diet for the remaining 20 months of their lives had an average daily dose of alachlor which was one-fifth of that fed to the group daily for 25 months. The incidence of tumours in the 5- to 6-month group, however (59% in males), was substantially higher than one-fifth of the incidence for the 25-month group (69% in males). In the absence of a proven threshold of safety, HPB postulated a linear model of dose-response which related tumour incidence to single daily rather than total lifetime dose. On this model, one exposure could be sufficient to cause cancer. Although inconsistent with the short-term mutagenicity studies, this assumption was, in the Branch's view, "biologically plausible and statistically consistent with the experimental data" (*Hearings*, p. 3676).

These assumptions were part of what HPB considered a "best-case scenario" (i.e., one which risked underestimating actual exposure), since (1) the patch tests used a small sample and the subjects tested were perhaps more careful than actual applicators would be, and (2) there could be additional exposure—from transportation, equipment maintenance, spills, clothing residues, etc.—which was not taken into account (*Report*, p. 58). However, faced with major uncertainty in the three key assumptions (HPB-E2 to E4), the Branch took a "prudent" or risk aversive stance on each, defended as reasonable in the context of health protection.

In summary, HPB interpreted the data on carcinogenicity and exposure submitted by Monsanto in a policy framework that elevated the value of health protection. The Branch thus chose a less rigorous criterion of causality, "cause for concern," to link rat tumours with alachlor exposure and in almost every instance interpreted uncertainty in the direction of the higher risk within the uncertainty range. While this interpretation could not be defended on the basis of the statistics alone, HPB argued that its analysis presented a reasonable response to the severity and kind of potential risk, and that the onus of proof was on Monsanto to counter with a scientifically compelling proof of safety.

Monsanto's Case for the Acceptability of Alachlor's Risks

In its petition to appeal the Minister's cancellation order Monsanto cited nine points of disagreement with the former's risk management decision. At the Alachlor Review Board Hearings Monsanto presented testimony and evidence in support of its position on all of these. We summarize here the disagreements that pertain to the risk estimation upon which the government based its decision.

The Issue of Risk Assessment Methodology

Monsanto's position on the issue of safety—that Risk-Benefit Analysis was the only fair way for regulatory agencies to approach the matter—relied implicitly on the availability of some method of quantifying both the risks and the benefits. On the benefit side, Monsanto used primarily economic measures (crop yields and prices). On the risk side, Monsanto recommended quantitative risk assessment. But as both the government and the Review Board pointed out, the paucity and uncertainty of the data in the risk and benefit studies submitted by Monsanto did not permit anything approaching the kind of scientific rigour promised by the quantitative approach. In anticipation of this response, Monsanto adopted a "backup" strategy of arguing that a "weight of evidence" or "qualitative" approach to the data would also demonstrate no unreasonable risks associated with alachlor use (*Hearings*, pp. 414, 420).

Regardless of the approach used, quantitative risk assessment or "weight of evidence," the implications for the Risk-Benefit Analysis were the same. The benefits side of the equation was fairly clearly established and quantified. On the risk side, however, because of the uncertainties and the meagre data base provided by the risk studies, it was impossible accurately to quantify the risks. When the uncertainties in the data base are interpreted in such a way that the risks are *minimized* (because there is no "weight of evidence" to the contrary), the invocation of Risk-Benefit Analysis is likely to yield the conclusion that the benefits outweigh the risks. When Monsanto recommended Risk-Benefit Analysis, it cannot have been blind to this implication of adopting the recommendation. We develop this point at greater length in Chapter V.

Carcinogenicity of Alachlor

Monsanto's position was that its long-term feeding studies on rats gave no reason to believe that alachlor posed a risk of cancer to human applicators of the chemical. The company admitted that the studies established the carcinogenicity of alachlor at relatively high doses in rats, but held (a) that there was no reason to conclude from the carcinogenic response in the rats that alachlor posed a similar risk to humans, and (b) that even if extrapolation from rats to humans were reasonable, the rat studies demonstrated a threshold of carcinogenic activity below which exposures produced no observed tumours and were, therefore, perfectly safe for humans. Thus, Monsanto argued that at the levels of exposure to be expected among human applicators and others—well below those which produced cancer in rats—alachlor was safe.

There is a tension between these two sides of Monsanto's argument which was noted neither by the Review Board nor by the other parties in the alachlor debate. The tension lies in the fact that if the first side of the argument, (a), is valid, then the best that can be said about the carcinogenicity of alachlor in humans is that it is a completely open question whether alachlor is carcinogenic in humans at *lower* levels of exposure than for rats, carcinogenic at *higher* levels, or even not carcinogenic in humans at all. The suggestion of argument (a) is clearly that the rat studies tell us *nothing* about the risks of alachlor. It thus serves Monsanto's purpose only on the assumption that the onus of proof lies with the government to prove risk before it cancels a product.

But the PCPA places upon Monsanto the responsibility to submit evidence to the government that supports its claims for the safety of its product. Thus it should not want to make its case on the grounds that its own studies are irrelevant to the issue. Argument (b) represents Monsanto's attempt to use its own studies positively, but at the expense, of course, of argument (a). Nevertheless, Monsanto relied upon both arguments throughout its presentation to the Board, depending upon which one served its purpose for the issue at hand. For example, argument (b) was used to establish a risk threshold in rats and to transfer this same threshold to humans to support a conclusion of zero human risk.[26] Yet, the metabolism studies submitted by Monsanto, ranging over the rat, mouse, monkey and human being, were held to show that rats were poor models of carcinogenic response in humans—argument (a). But obviously Monsanto cannot have it both ways. It cannot consistently maintain that the rat studies tell us *nothing* about the risks of alachlor for humans *and* that they tell us that, at expected levels of exposure, alachlor is *safe* for humans. Either we can extrapolate from rats to humans or we cannot. If we cannot, then Monsanto did not meet the regulatory requirement of submitting data to enable the government to determine that alachlor is safe. If we can, then Monsanto must enter the murky arena of debating whether the government's interpretation of the studies is acceptable.

Argument (b) is internally inconsistent as well, in that it depends on both a rigorous and a less rigorous interpretation of the data. Monsanto employed a rigorous criterion of proof to argue against the HPB conclusion that the data from low exposure levels in rats supported a causal relationship between alachlor and tumours. The company then shifted its approach to use a less rigorous criterion to conclude, not that the data were indeterminate, but that they were determinate with respect to the opposite conclusion, namely that the relationship was definitively absent and that a threshold of safety existed. The company's argument was based upon placing confidence in the finding that there were no significant increases in the incidence of tumours in rats at the lowest levels of exposure tested. While this finding could be simply a reflection of the relatively small number of rats tested, the belief of HPB officials, Monsanto chose to take it as conclusive and to use it as the basis for determining a quantitative threshold. As one Monsanto witness explained, if one finds a "No Observed Adverse Effect Level" (NOEL), one can divide by a "safety factor" (of, say, 100 or 1000) and thus reach a *zero probability* of risk (*Hearings*, p. 2617). If exposure levels are below such a threshold, one can be sure not just that there is an *acceptable* risk, but that there is *no risk at all*. This is what Monsanto claimed for alachlor in the Hearings: "We believe that carcinogenic or oncogenic risks to humans [from alachlor] don't exist. . . . There is no reasonable probability that alachlor presents a risk to humans" (*Hearings*, pp. 417, 420).

Carcinogenicity of Metolachlor

Monsanto held that the government had not only over-stated the risks of alachlor, but had also grossly under-stated the risks of its chief alternative, metolachlor. Monsanto produced confidential studies of other chloracetanilide compounds which, according to the *Report*, gave evidence of tumour reactions similar to those produced by alachlor. Even though higher doses of metolachlor than of alachlor were required to produce oncogenic response, Monsanto argued that the differences in the studies (different strain of rats, different experimental protocol) meant that the only firm conclusion to be drawn was that metolachlor, like alachlor, was an animal carcinogen. It could not be concluded, however, that it was a *less potent* carcinogen than alachlor (*Hearings*, pp. 5218ff). Here Monsanto appears to have used uncertainty as a basis for a conclusion of *increased* risk, rather than of decreased risk as in the case of alachlor.

Monsanto also maintained that well-water studies done in areas where both herbicides were used revealed that metolachlor was present in as high levels as alachlor and that metolachlor stays in the environment longer. In Monsanto's opinion this, in addition to the toxicological studies on metolachlor, undermined the government's claim that metolachlor was a safer, or even a completely safe, alternative to alachlor as an agricultural pesticide.

Estimation of Exposure: Determining the Margin of Safety

Monsanto argued that the studies they submitted on levels of applicator exposure to alachlor showed these levels to be far lower than those estimated by the government and also well below the demonstrated threshold of oncogenic effect. The company's interpretation of these studies, as of the rat and mouse studies, used both relaxed and rigorous criteria of proof, depending upon how these tended to reduce the estimation of alachlor's risk, thus reflecting a more risk taking attitude.

Most of the studies of applicator exposure were conducted after Monsanto submitted its initial studies to HPB, and thus represented data unavailable to the Branch at the time of the cancellation. The company carried out both patch and biomonitoring studies in 1984 and 1985. Although the patch tests showed levels from 20 to 240 times higher than the biomonitoring studies (depending on the assumed dermal absorption rate, 8.5-100%), Monsanto chose to base its exposure calculations only on the biomonitoring data. The company began its biomonitoring studies in 1984, collecting all urine over the 120-hour period following exposure, from 12 selected applicators in its employ. Applicators were protectively clothed and gloved; four used an open cab application system and the remaining eight, a closed cab. The following year, the company carried out 36 more urine analyses on applicators in closed cabs. To estimate total body dose from that recovered in the applicator's urine Monsanto assumed that the amount found in the urine was 88% of the total body dose. Generated from radioactive tracings of alachlor in rhesus monkeys, this value accounted for the amount of alachlor which was retained within the body.

In contrast to HPB's assumptions, the company argued the following:

Mon-E1. That biomonitoring was the state of the art procedure and provided a much more accurate indicator of total body dose than patch test results (*Hearings*, pp. 1438-40).

Mon-E2. That the company ought to be given the benefit of the doubt on applicators' adherence to the directions on the product label. Applicators ought, in other words, to be assumed to wear gloves and full protective clothing.

Mon-E3. That if patch tests were used, the results of dermal absorption studies on rhesus monkeys ought to be taken as valid for humans. Monsanto introduced new studies to the Board which showed a range of absorption from 2.7-9.4% (not taking into account one animal with a rate of 21.5%).

Mon-E4. That a threshold of safety or NOEL was demonstrated by the 25-month studies. The tumours produced in the 5- to 6-month rat study were therefore argued to be the result, not of the average daily, but of the total accumulated dose—which was higher in the 5-6 than in the 25-month study. Since the lower dose produced cancer in the nasal turbinates, they argued that it was not surprising that a higher dose, even though administered over 1/5 the time period, also produced cancers. Monsanto thus argued for full amortization, the calculation of "an equivalent average daily level of exposure from the intermittent exposures that a typical applicator would receive" (*Report*, p. 61).

Mon-E5. That spills and accidents were not part of "normal usage" and were thus to be ignored in the risk assessment.

Each of Monsanto's assumptions worked to decrease the calculated risk of alachlor's carcinogenicity. During the Hearings many problems with the company's assumptions and study design were pointed out by various witnesses. We cite some of these with our own comments:

E1. Biomonitoring studies are only more accurate than patch studies when supported by "adequate and appropriate metabolism studies" (US EPA Reinhart, 1986; in *Report*, p. 83) which are not available for alachlor. Moreover, urinalysis measures only the product excreted, not that lost during storage or breakdown into products which do not register in analysis. Monsanto witnesses attempted to allay concern that we have inadequate information concerning the way the human body metabolizes alachlor. Nevertheless, there is inadequate scientific foundation for the view that monkey metabolism is sufficiently similar to that in humans to allow the direct transfer of a correction factor for the urinary excretion value. Asked to choose between the two methods if the product's pharmokinetics were not known, Dr. Cowell, witness for Monsanto, stated, "I would have to use the patch test" (*Hearings*, p. 1418).

E2. All tests were conducted on Monsanto employees, some of whom were chemists with a high level of knowledge about the action of the product. The practices of these individuals may not have been typical of farm applicators; results may therefore underestimate exposure (*Report*, p. 81, *Hearings*, pp. 1400-04). In fact, the only information available on the use and effectiveness of protective clothing by farm applicators was anecdotal and included statements that some farmers used no protective clothing and no gloves. Monsanto thus assumed a type of protection in its scientific study to which there were empirically established counter-instances. Nevertheless, Monsanto felt this was procedurally fair to the company as a regulatory assumption. Similarly, Monsanto's estimates for commercial applicators assume a closed application system and full protective clothing, assumptions which, as above, were considered fair for regulatory purposes, though possibly empirically inaccurate.

E3. As discussed above, Monsanto's arguments on the determinacy of animal data for humans contradict one another. The company argued both that animal data should be regarded as indeterminate on issues of human health and that selective statistics taken from animal test results should be used without translation in the calculation of risk to human health. Although argument Mon-E3 does not feature in the defence of its own risk assessment, which is based upon biomonitoring data, it does demonstrate the flexibility of the company's principles.

E4. The assumption of a threshold of safety on such sparse data reflects a strong risk taking stance. However, Monsanto went further than this. Its amortized estimates assumed exposure for only 1 day per year for farmers and 15 days per year for commercial applicators. By contrast, the Board's initial report on the request for temporary registration of alachlor accepted 6 and 30, respectively (*Report*, Appendix, Table B).

E5. It was admitted by Monsanto witnesses during the Hearings that not only do accidents occur on farms, but alachlor was accidentally spilled during one of the biomonitoring experiments (*Hearings*, p. 1410; *Report*, p. 79). Comment E2 applies here.

It can be seen that the data Monsanto submitted to the government and the Review Board were incomplete and open to some question. In order to interpret these data as actually demonstrating product safety, the company was required to work strategically with the uncertainties in the data, making whichever assumptions were necessary at each juncture to reduce the calculated risk of alachlor, while increasing the calculated risk of metolachlor. The company defended its interpretation as reasonable in the context of product regulation, where, in its view, any bias ought to favour fairness to the regulated and to minimize benefit loss to the community at large.

Monsanto's risk estimation demonstrates two key features: (1) that risk assessment is necessarily motivated and guided by an underlying framework of values, which influences the assessment by framing the process in a distinctive way, and (2) that while pursuit of these values is not necessarily inconsistent with maintenance of scientific integrity, where there are substantial uncertainties in the science they readily lend themselves to strategic manipulation in a regulatory context.

Burden of Proof

There was a major point of difference between Monsanto and the government which was not explicitly articulated by either. This relates to the issue of who ought to shoulder the burden of proof. As noted earlier, the government maintained that Monsanto bore the burden of proof. By contrast, Monsanto's position was that this burden rested firmly upon the government: for the government to intervene in the marketplace by excluding one product while accepting that of its competitor, it must have clear, nonarbitrary and noncapricious reasons for doing so. This means that the government not only must be certain of the grounds upon which it finds one product *unsafe*, but it must also be certain that the competitive products on the market are *safe*, or at least *safer*. In the debate before the Alachlor Review Board, Monsanto cast this issue as one of "fairness"; thus it claimed that the central issue before the Board was "fairness" rather than "safety" (*Hearings*, pp. 264, 290-91).

For Monsanto there was a fundamental unfairness involved in the cancellation of alachlor on the grounds of the mere *possibility* of human carcinogenicity, based upon uncertain evidence, while the registration of the competitor product, metolachlor, was continued despite equally uncertain evidence of its safety. To be sure, the evidence of metolachlor's oncogenicity was not as strong as that for alachlor. Nevertheless, the government had not provided a scientifically compelling case for the view that it was not in fact as risky as alachlor. Thus, from Monsanto's perspective, the only way to avoid arbitrariness was to require the government to bear the burden of proof—to demonstrate alachlor's risk to humans *as well as* the relative safety of the competition.

Of course, as previously noted, the incompleteness of the data provided to the government by the producers of both alachlor and metolachlor made it impossible to demonstrate statistically either that alachlor posed clear risks to applicators and others, or that metolachlor did not pose such risks. Monsanto believed that the statistical proof available and the "weight of (remaining) evidence" combined to support the conclusion that alachlor did not pose unacceptable risks to humans, and that it was up to the government to come up with statistical evidence to the contrary. The government, however, took the view that there was sufficient evidence of alachlor's risks to warrant exercising caution on the side of safety. Monsanto witnesses viewed this as a case of considering the product guilty until proven innocent, and held, in effect, that since there is no way to *prove* safety, there was no way for a producer to meet the onus of proof the government required (*Hearings*, p. 1066). The government appeared, however, to require not so much *proof* of safety as evidence more heavily *weighted* in favour of safety, an onus it believed Monsanto had not met. Monsanto held that the government bore the onus to put forward the weighty evidence on the other side. The problem was that none of the evidence concerning alachlor's impact on humans was very weighty, and neither party accepted that *it* had the responsibility to provide more or better evidence.

In this chapter we have examined the case put forward by the government in support of its conclusion that alachlor presented a risk to human health high enough to warrant its exclusion from the Canadian market, and the case put forward on the opposite side by the producer of alachlor that the risk was many orders of magnitude smaller than the government claimed. Both sides, it must be remembered, came to their respective conclusions on the basis of essentially the same data, though each supplemented those data with additional data in support of their interpretations. Both claimed to be doing scientifically objective assessments of the risk, though both recognized that the "regulatory science" involved here was not able to meet the high standards of a rigorous, quantitative science. We have identified the way in which several critical assumptions significantly influenced the divergent assessments of risk offered by the two parties, and we have argued that these assumptions are motivated importantly by very different value frameworks having to do with attitudes toward the appropriate liberties and responsibilities of citzens and government as well as attitudes towards risk itself.

The Alachlor Review Board was called upon to adjudicate between these two assessments of risk, as well as between the two conclusions about the proper management of those risks. It, too, interpreted its task as the impartial arbitrator, guided as much as possible by scientific objectivity. In Chapter IV we will consider how the value framework of the Review Board worked to shape its assessment of alachlor's risks.

IV

The Alachlor Review Board's Estimation of Alachlor's Risks

The Alachlor Review Board found considerably more merit in the arguments put forward by Monsanto than it did in those put forward by the government. It concluded that the government had overestimated the risk of alachlor and underestimated the risk of metolachlor, reasoning that, although both products are "potential human carcinogens," the risks they pose to human beings at expected levels of use are negligible or nearly so. This finding of minimal risk, of course, was the main support for the final conclusion that alachlor should be registered for sale in Canada. The conclusion was further supported by a determination that "the continued availability of [either alachlor or metolachlor] is essential if corn and soybean production in Canada is to remain economically viable and internationally competitive" (*Report*, p. 6).

In reviewing the Board's reasoning in arriving at these findings, we shall have occasion to point out a number of apparent contradictions, questionable assumptions, and simple oversights. The object of the exercise, however, is not to criticize the Board and thereby defend the government's original decision to cancel the registration of alachlor. The Board's conclusions, and the arguments in their support, are of interest to this particular study only insofar as they tell us something about the risk assessment process. This dictates a certain selectivity on our part: the arguments and conclusions to which we shall call attention are those that exhibit not just the pres-

ence but the inevitability of value assumptions in the Board's estimate of alachlor's risks.

These assumptions will be spelled out in some detail in Chapters
V and VI below. At one level, as discussed above, they concern the
burden of proof and the values that are primarily at stake when the
issue, whether the risks posed by alachlor are acceptable, is confronted. We show below that the Board was more sensitive than the
government to the potential economic benefits of continued alachlor
use, and less sensitive to the health benefits associated with cancelling its registration. At another level, these assumptions concern the
nature of rationality, the place of technology in our lives, and the role
of the state in regulating private individuals and activities in the
"private sector."

With respect to these matters, we point out that the Board held to
the familiar conception of "instrumental rationality," a view which,
in the context of technology debates, is fairly narrow and contentious. Secondly, the Board exhibited an optimistic attitude toward
the potentiality of technology for solving human problems. This
optimism was evident in the Board's dismissal of the suggestions
that the registration of both alachlor and metolachlor should be cancelled and that stringent controls should be placed on their use, and
more generally in its tacit endorsement of unrestricted "chemical
farming." Thirdly, the Board took the standard "liberal" view that
coercive governmental measures—in this case, regulation of producers of chemical products used in agriculture—are justified only if
there is *compelling* evidence that the measures are necessary to protect an important public interest—in this case, human health. In
terms of the model described in Chapter II, these higher level
assumptions form a value framework that leads the risk assessor to
adopt the lower level assumptions. The latter assumptions then control the risk estimation by fixing a frame within which the scientific
data are interpreted. As indicated there, a decisive component of the
frame is the assumption concerning burden of proof.

To the reader of the Board's *Report*, the first inkling that it had
placed the burden of proof on the government comes when it is
noticed that the Board consistently focused on the credibility of the
government's estimation of the risks of alachlor and metolachlor and
tended to reject government conclusions that could not be fully supported. In one sense, this is to be expected: Monsanto was appealing
a government ruling and thus was calling into question its reasonableness; so the challenge to the government, in defending itself
against the appeal, was to show that its decision was, in fact, reasonable. But in laying a burden of proof on the government the Board
was imposing a more stringent requirement. It was requiring the

government to produce a scientifically compelling case for the position that the risks posed by alachlor were unacceptable.

We have seen in Chapter II that this demand is different from and more stringent than the requirement that the decision to cancel alachlor's registration merely be reasonable. This can be seen operating specifically here in the following way. The government's case for cancellation was, in part, that there is some evidence to show that alachlor poses a threat to human health and when public health is at issue it is better to err on the side of caution, by minimizing the threat. But although this is arguably a reasonable line to take in defending the cancellation decision, as we have already pointed out, it does not rest on a scientifically compelling case for the particular estimate of alachlor's risks at which HPB arrived. Priorizing the health issue justifies the assumption that, if err one must, it is reasonable to err on the side of caution. By demanding of the government that it provide scientifically compelling evidence for its estimate, the Board unwittingly rejected the government's priorities (and, as will be seen below, priorized economic benefits instead). The reason is that the demand in effect denied to the government the option of "erring" on the side of the particular values it had an interest in protecting. And given the massive uncertainties met in conducting the risk assessment, the demand to "prove" its case posed a challenge that the government probably would not have been able to rise to, even if those who defended its position had been considerably more persuasive than were the individuals who appeared at the Hearings on the government's behalf.

In the Board's perception, HPB had seriously over-estimated the risks of alachlor. It had done so by making risk-aversive or conservative (what we shall later term "risk-cautious") assumptions at numerous critical points where there were uncertainties in the risk estimation.[27] We have seen how this conservatism showed up in HPB's claims that the adenomas found in the rats' nasal turbinates ought to be included in the risk assessment as precursors of adenocarcinomas; that the statistically non-significant stomach carcinomas at low doses ought to be considered part of the dose-response pattern and hence treated as *biologically* significant; and that the mouse oncogenicity study established that alachlor caused cancer in mice as well as rats. Similar conservatism, the Board found, was central to HPB's exposure estimates. These estimates rested on the assumptions that applicators would wear inadequate protective clothing; that 100% of the product deposited on the skin would be absorbed into the blood stream; and that an open application system would be used. In the Board's view, the cumulative effect of conservative assumptions concerning the animal feeding studies was to suggest

considerably greater carcinogenicity of alachlor in rats and mice than
the data warranted. The cumulative effect of HPB's conservative
assumptions concerning applicator exposure was that it constituted
an unreasonable worst-case exposure scenario.

The Review Board questioned these conservative assumptions
during the Hearings and rejected many of them in its own estimate.
It rejected the conclusions that the mouse oncogenicity study established that alachlor causes cancer in mice, and thus reduced HPB's
finding of carcinogenicity in two animal species to one. And more
importantly, with respect to the estimation of applicator exposure
the Board concluded that biomonitoring gave a better reading of
alachlor absorption levels than did the patch tests used by HPB, even
though the patch tests gave higher readings and the metabolic information necessary to validate the biomonitoring results was unavailable; that it was appropriate to amortize exposure levels (though not
to the extent done so by Monsanto); and that in estimating exposure
levels it was appropriate to assume that the applicators would wear
full protective clothing and would employ a "closed application system" (closed cabs and bulk tanks). All of these revisions of HPB's
assumptions contributed to an estimate of applicator exposure
dramatically lower than HPB's estimate. Where HPB's assumptions
yielded an exposure estimate at the same level as the lowest level at
which a tumour was observed in the rat feeding studies, the Board's
revised exposure estimate suggested that applicator exposure would
be three to four orders of magnitude lower than the lowest level at
which a tumour was observed in those studies.

In its estimation of the risks of alachlor, the Board basically saw
itself as counter-acting HPB's excessive "regulatory prudence" by
careful interjection of "scientific prudence." It thought this necessary
because a decision to cancel registration of alachlor based on an
unreasonable worst-case exposure scenario would lead to "unjustified costs and unjustified results." The unjustified costs the Board
had in mind were the foregone benefits for the farm community
associated with alachlor use; the unjustified results, the unfairness to
Monsanto of denying to it the freedom to market its product.

The Board thus appears to have seen that HPB's estimation of
alachlor's risks was guided by its desire to protect a salient value,
public health. But the Board interpreted this as leading to an *over-
estimation* of alachlor's risks. It did not appear to recognize that it too
was guided by values it saw as salient—the economic benefits associated with availability of alachlor and the freedom of the manufacturer, Monsanto, to market its product in a fair competition with
other manufacturers' products—and that its own value framework
also operated in the risk assessment by framing the process in a dis-

tinctive way. Rather, it appears to have conceived the scientific prudence it brought to the risk estimation of alachlor as a value-free and
socially neutral procedure that had the effect of pruning away HPB's
excesses to reveal the *actual* risks of alachlor. In our view, it did not
see that the choice between the use of "liberal science" and "conservative science" for regulatory purposes is itself a choice expressive
of a commitment to the priority of values such as safety, on the one
hand, and economic benefit, market fairness, etc., on the other.

It is significant that the Board explained HPB's use of the term
"conservative" in this way: " 'Conservative' was used by HPB in the
sense of an estimate that overestimates the actual exposure" (*Report*,
pp. 58ff). It is not likely that HPB officials would have accepted this
account. Surely their sense of a "conservative" estimate had more to
do with situations where there is uncertainty or indeterminacy in the
data which precludes rigorously scientific conclusions, and where
the estimators recognize a range of possible exposure and choose to
adopt the higher end of the range because they prefer to err on the
side of over- rather than under-estimating the exposure *if they turn
out to be in error*. This is something quite different from a deliberate
over-estimating of a known level of exposure.

The Board's interpretation of conservatism supposes that there is
an objectively correct estimate of the exposure, which can be
discovered by being scientifically rigorous, using rigorous criteria of
causality. It implies that what the Board saw as HPB's "conservatism" led to a *mistake*, an *over*-estimation of the actual exposure to
alachlor. But this ignores the significance of the uncertainty. When
there is uncertainty, a *correct* estimate cannot be given, or cannot be
known to have been given. There is no benchmark level of actual
risk, as an objective fact, with reference to which an estimate of risk
can be characterized as an under- or over-estimation. The uncertainty creates a need to go beyond the evidence. The alternatives
presented by the uncertainty are either to give up, an option not
available to HPB or the Board, or to make a judgement that "risks"
over- or under-estimating the exposure. To repeat a point made in
the preceding chapters, a judgement that *risks* being an over-estimation of exposure is not the same as an over-estimation. In the context
of uncertainty, one cannot know one is over-estimating. But one can
know that, in the circumstances, in view of what is at stake, it would
be better to be "guilty" of over- than of under-estimating.

Put otherwise, the question the estimator, and particularly the
regulator, faces is, Which shall it be? Shall I risk erring on the high
side or the low side? The decision to risk erring on the high side (if
this means representing the exposure as higher rather than lower
than it might be represented as being) is not a decision to over-esti-

mate the exposure. The desire is (or ought to be) to give the *best* estimate. But in the face of uncertainty the best estimate can only be
identified as "best" by appeal to the appropriate values at issue.
That is, "best estimate," in the context of uncertainty, is value-rela-
tive.

The Board concluded that there was a three to four orders of magnitude gap between the estimated applicator exposure in what it
termed a "reasonable worst-case exposure scenario" (its own estimate) and the lowest dose at which a tumour was observed in one of
the rat studies, and that this represents a "reasonable margin of
safety." The Board's conclusion and, indeed, its own definition of the
"reasonable worst-case exposure scenario" *are* reasonable only from
the perspective of a particular view of the values at stake.

To this point, we have only stated our view of the inevitability of
value assumptions in risk estimation, and have traced the outlines of
an abstract line of reasoning that supports the view and underlies
the model described in Chapter II. The remainder of this account of
the Board's estimation of alachlor's risks will further clarify the view
and indicate the ways in which it provides the best perspective for
understanding the Board's investigation. It is especially in our analysis of the Board's conclusions concerning a "reasonable worst-case
exposure scenario" and a "reasonable margin of safety" that the
view and model are confirmed.

The Carcinogenicity of Alachlor

On the basis of the rat feeding studies, the Board determined that
"alachlor is an animal carcinogen and should be considered a potential human carcinogen" (*Report*, p. 7). Choice of the term "potential"
was significant. The Board held that the IARC classification of
alachlor as a "probable" carcinogen, which HPB accepted, was misleading because it suggested a probability greater than 50% that
alachlor would induce cancer in humans. The Board preferred to say
that it was a "potential" carcinogen, since this was "devoid of any
quantitative implications" (*Report*, p. 61).

But although the Board concluded that alachlor is a potential
human carcinogen, it did not regard it as likely that alachlor would
actually induce cancer in humans. Its estimate of alachlor's risks
divides naturally into two parts. The first part is an interpretation of
the rat and mouse feeding studies. The second part is an application
of that interpretation to the issue of alachlor's potential as a human
carcinogen. Since the humans primarily at risk were assumed to be
the applicators of alachlor, in this second part of the estimation the
Board was mainly concerned to quantify applicator exposure. It tried

to arrive at a "reasonable worst-case exposure scenario." As noted above, the critical conclusion was that a reasonable worst-case exposure scenario indicates that, in the worst case, applicators would be exposed to a quantity of alachlor that is "1,000 to 10,000 times (three to four orders of magnitude) lower than the lowest dose at which a tumour was observed in one of the rat studies. In this situation, the Board considers this to be a reasonable margin of safety" (*Report*, p. 84).

The burden of the Board's argument is borne by its working out of an exposure scenario, which constitutes the second part of the assessment. Before turning to that, however, some attention must be paid to the animal feeding studies. The Review Board took an interpretive approach in its analysis of these studies, which it understood as scientifically rigorous and value-neutral. That the difference between its approach and that taken by HPB might be a function of values which each sought to protect did not seem to occur to Board members. This is an issue which we will take up in some depth below. It will be recalled that HPB's approach to these studies was to weigh the evidence overall for both alachlor and metolachlor and, on the basis of features such as dose-response relationship, to assess the relative and absolute carcinogenicity of the two chemicals. The Review Board, focused as it was in these matters on the need to remain scientifically credible in the face of massive uncertainty, did not share the view that a comparative analysis was possible. According to the Board, the variation in test conditions (e.g., different strains of rats, different laboratories, different protocols) made a scientifically reliable comparison between the two data sets impossible (*Report*, p. 70). The Review Board therefore read the alachlor data base for confirmation of carcinogenicity only; it did not build a case for the relative carcinogenicity of alachlor and metolachlor.

We have already cited the numerous problems the Board found with the government's interpretation of these studies and no doubt its reinterpretation of them increased its confidence in the "three to four orders of magnitude" claim. In points through HPB-A5, we noted the critical judgements made by HPB concerning the data presented by the rat feeding studies. Parallel with these, we note the following features of the Board's estimate:

ARB-A1. The Board made no mention of the rarity of nasal tumours.

ARB-A2. The Board did not dispute HPB's conclusion that the tumour data at both nasal turbinate and stomach sites exhibited a dose-response relationship (*Report*, p. 61).

ARB-A3. In finding a dose-response relationship at the nasal turbinate site, the Board did make a connection between the adenomas and the carcinomas.

ARB-A4. In finding a dose-response relationship at the nasal and stomach sites, the Board also interpreted as biologically significant the nasal and stomach tumours occurring at low doses.

ARB-A5. The Board accepted the opinion that the brain tumours originated in the nasal turbinates, but concluded that this evidence showed alachlor to cause cancer at two rather than three sites; it did not interpret the metastases of the nasal tumours as evidence of alachlor's carcinogenic potency (*Report*, pp. 75, 93).

ARB-A6. The Board made no mention of the range of tissue affected or the number of tumours produced.

The effect of the Board's doubt that the animal test data could establish relative carcinogenicity in humans was thus to find in the rat data confirmation only of alachlor's carcinogenicity and not of its potency. Its further conservatism in the application of these test results to humans led the Board to conclude that while alachlor was demonstrably an animal carcinogen, it should be regarded only as a "potential human carcinogen" (*Report*, p. 7). This judgement appears to have reduced the concern over the chemical originally expressed by HPB's Toxicology Evaluation Division in its 1982 report: "The recently submitted replacement studies for chronic rat and mouse IBT studies indicate that Alachlor is one of the most potent carcinogenic pesticides currently in use" (*Report*, p. 34).

The Board's view of the mouse oncogenicity study was that it was inconclusive, a view which led it to reject the study's relevance to the question at hand, but from which it concluded nonetheless that "alachlor should be considered to be carcinogenic in only one species of animal, the rat" (*Report*, p. 65).

This dismissal of the government's interpretation of the mouse oncogenicity study is important because it indicates something about the Board's way of dealing with uncertainty. HPB's interpretation "was challenged by Monsanto at the Board's hearings on the basis that the apparent statistical significance was caused by an abnormally low incidence of spontaneous tumours in the control group of female mice" (*Report*, p. 64). After considering the evidence, the Board concluded that "there is no convincing evidence that alachlor induced tumours in this mouse study" (*Report*, p. 64).

The Board offered five considerations in support of this conclusion. These go some distance toward establishing the Board's view, but it cannot be claimed that they are conclusive. For the most part, they do not establish that the lung adenomas and adenocarcinomas were *not* treatment related, which is the point at issue, but only call into serious question the claim that they *are* treatment related. The considerations were (1) that the incidences of lung adenomas in the control and treated male mice and in the treated female mice were the same, and it was only the female control group that showed lower incidence; (2) if the tumours were treatment related the Board would expect to find (but didn't find) increased multiplicity of lung adenomas in the treated mice; (3) "carcinogens usually cause increased incidence of tumours at shorter times of exposure as compared to controls. This is not seen in this bioassay with alachlor" (*Report*, p. 65); (4) there was no change in the ratio of benign to malignant tumours, as would be expected if the tumours were treatment related; and (5) no tumours or neoplasms were found in other organs or tissues, as would be expected if the tumours had been induced by alachlor.

It will be noted that throughout these five considerations there is considerable uncertainty. One cannot say with any great confidence that absence of increased multiplicity of lung adenomas or absence of change in the ratio of benign to malignant tumours is a clear indication that the tumours were not treatment related. These are considerations that count *against* the claim that the tumours are treatment related; their significance is that they weaken the claim. It is consistent with acknowledging their force to find that equally strong considerations can be brought against the counter-claim that the tumours were *not* treatment related.

Similar observations could be made concerning much of the reasoning employed by the Board in reducing HPB's estimate of alachlor's risks. It is often necessary and generally reasonable in argumentation to employ considerations of the sort the Board employed in interpreting the mouse oncogenicity study. But, given the context of uncertainty, it is dangerous to reason in this way without due sensitivity for the risks associated with being wrong, that is, for the values that will be sacrificed. And, in any case, to reason in this way, by citing considerations that weaken an opponent's argument, and then on the basis of having weakened it to infer that the opponent's conclusions are false, is to impose on the opponent a strong burden of proof.

We shall reserve discussion of the Board's estimate of applicator exposure until after considering its estimation of the carcinogenicity of metolachlor. As will be noted, the Board's argument for the view that the risks of metolachlor are on a par with those of alachlor involves an attribution of the burden of proof to HPB, and in this way parallels its argument against HPB's interpretation of the mouse oncogenicity study considered immediately above. But there is an interesting twist. The effect of its objection to HPB's interpretation of the mouse oncogenicity study was to find that HPB had over-

estimated the risk (of alachlor), presumably in the interest of public
health. By contrast, the effect of its objection to HPB's view that
metolachlor is safer than alachlor, similarly based on placing a burden of proof on HPB, was to find that HPB had *under*-estimated the
risk (of metolachlor). On the surface, at least, the Board was taxing
HPB with having been overly concerned about human health in its
estimation of alachlor's risks, and insufficiently concerned about
human health in its estimation of metolachlor's risks. The Board may
well have been right about this. However, it is of interest that it did
not seem aware that it had simply reversed the error, based on an
opposite priorization of values. In effect, the Board was charging
HPB with failing to prove that alachlor is unsafe and that metolachlor is safe. From this, of course, it does not follow that both are
safe (or even that they pose similar risks). The Board's conclusion
makes sense only on the assumptions (1) that the government has
the burden of proof and (2) that the option of cancelling the registration of both herbicides is not a serious alternative. The net effect of
equating the risks of alachlor and metolachlor was to represent both
as safe, not both as unsafe. Thus, the implication is that (since the
idea of banning metolachlor was not even an option) the real function of its arguments for upgrading the risks of metolachlor was not
to assert concern for public health but, by putting the two herbicides
on a par, to ground the position that both should be available to Canadian farmers. We have already noted that this appeared to be
Monsanto's strategy in its argument to the Board that HPB had
under-estimated the risks of metolachlor, and the strategy appeared
to have worked well with the Board. Despite the apparent shift from
concern not to over-estimate alachlor's risk to concern not to underestimate metolachlor's, the Board's insistence that HPB bear the burden of proof had the consistent effect of undercutting HPB's position
that the risks of alachlor are greater than those of metolachlor. That
is to say, in the first case HPB was given the burden of producing the
hard scientific evidence that alachlor was unsafe, and in the second
case of producing similar evidence that metolachlor was safe. In neither instance could HPB meet the rigorous standard. The result of
undercutting HPB's position was to replace its conservative, risk-
aversive approach to the estimation of alachlor's risks by a risk-taking (or, as one might say, gambler's) approach. We discuss this
implication further in Chapter V.

The Carcinogenicity of Metolachlor

HPB had interpreted the two rat metolachlor feeding studies as not indicating carcinogenic activity in the strain of rats tested. The Board's interpretation of those studies differed radically:

> The first study showed combined incidents of liver nodules plus liver hepatocellular carcinoma ... of 11/60, 3/60, 1/58 and 1/54 in female Sprague-Dawley rats at 3,000, 1,000, 300, 30, and 0 ppm of metolachlor. The second study showed incidents of 7/60, 2/59, 1/60 and 0/59 in females at 3,000, 300, 30 and 0 ppm of metolachlor and 9/60, 2/60, 2/57, 2/60 in males. Thus, metolachlor in two independent studies showed induction of cancer and a cancer precursor in the liver. These data demonstrate a clear, dose related phenomenon. (*Report* p. 69)

In addition, on the Board's reading, the studies showed incidences of nasal turbinate carcinomas and the induction of two liver cancers of a different and uncommon type, angiosarcoma.

In interpreting the two studies, the Board made the following assumptions:

ARB-M1. "The relative rarity of [the nasal turbinate] tumours does not allow for the conclusion that they are a statistical anomaly" (*Report*, p. 69).

ARB-M2. The data concerning liver nodules and hepatocellular carcinomas indicate a clear dose-response relationship.

ARB-M3. The liver nodules and hepatocellular carcinomas should be considered together, since "there is considerable evidence, quite good evidence in some instances, that nodules can be a precursor for cancer in the liver" (*Hearings*, p. 3939).

ARB-M4. The rarity of the other type of liver cancer, angioscarcoma, suggests that its occurrence too is biologically significant (*Report*, p. 69).

It can be seen that the uncertainty surrounding each of these issues was read in favour of the possibility of carcinogenicity, a disposition towards risk augmentation that the Board did not exhibit in its assessment of alachlor.

So far we have considered the Board's view of the potency of metolachlor in rats. The extrapolation of these results to humans presents a distinct set of issues. We focus on the sceptical principle employed by the Board: "Even if the relative potency of two chemicals in an animal species is known, it cannot be assumed that the same relative potency will apply in humans" (*Report*, p. 71). This principle is enunciated in a context where it is accepted as a matter of regulatory prudence that if "a chemical is carcinogenic in an animal species [this] implies that it is a potential human carcinogen" (*Report*, p. 71). The latter principle, of regulatory prudence, gives the Board its reason for inferring that metolachlor (like alachlor) is a potential human carcinogen. The former, sceptical principle provides the basis for its refusing to draw inferences from the evidence about the relative carcinogenicity of the two chemicals. The Board thus refused to infer, from the evidence that higher doses of metolachlor than of alachlor are required to induce nasal cancer in rats, that metolachlor is less carcinogenic *in humans*. This inference, which HPB was willing to make, is blocked by the idea that extrapolation of similar relative potencies is unreliable unless one has evidence that animals and humans metabolize the chemicals in the same way. Finally, from its refusal to infer that metolachlor is safer than alachlor the Board concluded that "the prudent position to take is that [metolachlor and alachlor] should be considered equally hazardous to human health" (*Report*, p. 71).

It might be claimed that the discrepancies in the experimental procedures used for the two chemicals (e.g., different laboratories, different rat species, different grades of the chemical tested) gave the sceptical principle strong support. However, such considerations do not affect the logic (or illogic) of the argument. In effect, the reasoning follows this pattern: we cannot infer that such-and-such is the case; therefore, such-and-such is not the case. (We cannot infer that metolachlor is safer than alachlor for humans, despite the fact that higher doses of metolachlor than of alachlor are required to induce nasal cancer in rats; therefore, metolachlor and alachlor are equally safe for humans.)

Given that this argument is used to minimize the significance of the differences in dose-response relationship, it expresses a risk-taking rather than a conservative, risk-aversive approach to risk estimation. Remarkably, in stating its conclusion, the Board alludes to "prudence." It evidently thinks the position prudent because it serves to undermine Agriculture Canada's conclusion that metolachlor is "much safer" than alachlor. But this ignores the fact that the effect of the conclusion that the two herbicides are equally safe is to discount the apparent significance of the need to administer higher doses of metolachlor to induce cancer. That is, the real function of the conclusion is not, as the Board represents it, to maintain that metolachlor is less safe than one might think; it is to discount the impression that, because it takes less alachlor to induce nasal cancer, alachlor is more risky than metolachlor.

On the face of it, the Board's logic here is problematic. We conclude this section by asking: on what assumption might the Board be justified in reasoning as it does? If chemical A is less carcinogenic in animals than chemical B, in light of the "sceptical principle" we cannot infer that it is less carcinogenic in humans. We can only conclude that "*It is not possible to say* whether chemical A is less carcinogenic in humans than chemical B." In order to go on to the conclusion that "Chemical A and chemical B are *equally* carcinogenic," (or even, as the Board did, "equally noncarcinogenic") what must be assumed? The most straightforward assumption that would justify this inference is the following one: the party which maintains that chemical A and chemical B are *not* equally (non-)carcinogenic has a burden of proof; thus, if it cannot shoulder that burden, we ought to conclude, as a matter of prudence, that its view is mistaken, and hence, that the two chemicals *are* equally (non-)carcinogenic.

Attributing this assumption to the Board gives us an interpretation of its position regarding the carcinogenicity of metolachlor that fits with what we have concluded concerning its approach to other aspects of its risk assessment of alachlor. On this view, what it is claiming is that HPB needs to provide compelling evidence of the accuracy of its risk estimate, in this case, for its view that metolachlor is safer than alachlor. If HPB cannot make its case, then its view of the matter should be taken as false. And if HPB's view is false, then metolachlor and alachlor are equally safe. Of course, HPB cannot make its case without contradicting the sceptical principle that one cannot extrapolate relative potencies from animals to humans.

The Relative Carcinogenicity of Alachlor and Metolachlor

Discussion of the Board's estimation of alachlor's and metolachlor's carcinogenicity has touched on the assumptions underlying its argument for the view that alachlor and metolachlor are equally carcinogenic. Here we shall only single out a special feature of that argument, which confirms the analysis set out in the preceding paragraphs. It is noteworthy that the language of the Board's argument in the Report shifts. We meet sentences like the following: "*It cannot be concluded* from the available data that either substance is a more potent carcinogen than the other, even in rats" (*Report*, p. 70); "Any conclusion that one of the two chemicals is more or less potent than the other *is unwarranted*" (*Report*, p. 71); "Even if the relative potency of two chemicals in an animal species is known, *it cannot be assumed* that the same relative potency will apply in humans" (*Report*, p. 71); and most telling—"For regulatory purposes, alachlor and metolachlor should be considered to be more or less equivalent in carcinogenic potency from a human perspective, *since there are no data in humans to indicate otherwise"* (*Report,* p. 71; stress added throughout).

The significance of these turns of phrase is that the Board is finding the government's evidence for metolachlor's lower risk to be less than scientifically compelling, on this basis is rejecting it, and is using the rejection to conclude that alachlor and metolachlor are on a par with respect to safety. To appreciate the leap this conclusion involves, one need only reflect that there are no data to indicate *anything* about the relative carcinogenicity of these two chemicals in humans. In that case, as far as the data go, *all* possibilities remain open: that the two are equally carcinogenic, but also that metolachlor is more carcinogenic than alachlor, and that alachlor is more carcinogenic than metolachlor.

The Board then concludes that there was no valid scientific basis for the impression formed by Agriculture Canada that metolachlor was "much safer" for humans than alachlor. Unless exposure patterns are substantially different for the two chemicals, the prudent position to take is that they should be considered equally hazardous to human health (*Report,* pp. 71, 94).

We recognize that the Board had other reasons for equating the risks of metolachlor with those of alachlor. In particular, it saw the significance of the fact that "alachlor and metolachlor are of the same class of chemical compounds (chloracetanilides) and could be anticipated to invoke similar biological responses" (*Report,* p. 69). With this in mind, from the foregoing account of how the Board placed the burden of proof on the government it cannot be inferred that the Board was mistaken in concluding that alachlor and metolachlor are equally safe (or unsafe). But if they are on a par with respect to safety, and if HPB's risk-aversive or conservative approach to the estimation of alachlor's risks was appropriate, then the option that none of the parties to the alachlor debate took seriously, that of cancelling registration of both herbicides, or, to be more practical about it, placing serious restrictions on their use, deserves considerably more attention than it received.

The Argument for a Reasonable Worst-Case Exposure Scenario

As indicated in discussion of HPB's and Monsanto's estimations of the risks of alachlor, a critical step is the determination of the extent to which individuals, especially applicators, will be exposed to the herbicide. The centrepiece of an estimation of alachlor's risks will be a presentation of the gap, if any, between the dose sufficient to induce tumours in rats and the dose to which applicators are likely to be exposed. The central, bottom-line decision will be whether that gap provides a reasonable margin of safety. Reaching a view on this matter requires developing an exposure scenario, and it is also important that this scenario be reasonable. But since human health is at issue, all parties agreed that this scenario should provide for the reasonable *worst* case. They did not agree, however, about *which* was the reasonable worst case.

We have already seen that the critical elements of an applicator exposure scenario comprise at least four categories of assumptions: assumptions concerning the protective clothing, if any, applicators will wear; amortization assumptions; assumptions concerning the frequency of application of the herbicide and the quantity applied; and assumptions concerning the proper method for developing an estimate of the amount of the herbicide that will enter the applicator's system (patch tests or biomonitoring). Alongside each of the exposure estimates cited in Figure 1, we have indicated the assumptions under these four categories that result in the estimate. Exposure estimates vary by many orders of magnitude depending on what one decides with respect to each category of assumptions. Such estimates are most sensitive to assumptions concerning protective clothing and, especially, to amortization assumptions. As we shall see, all of these assumptions are, to use the terminology introduced in Chapter II, either inherently or conditionally normative.

The Board employed the following assumptions in developing its worst-case applicator exposure scenario:

ARB-E1. Biomonitoring is a more reliable method than patch tests for measuring exposure (*Report*, p. 83).

ARB-E2. When patch test data are used, a 25% dermal absorption rate should be assumed (*Report*, p. 76).

ARB-E3. Commercial applicators should be assumed to wear full protective clothing, including gloves (*Report*, p. 83). For non-commercial applicators, the values derived from protected applicators should be multiplied by a factor of 13 to account for cases where protective clothing is not worn or is not effective (*Report*, p. 81). Despite this recognition of unprotected farmers' increased exposure, however, the risk levels ultimately estimated by the Board were based on the assumption that farmers would wear effective protective clothing (see Figure 1).

ARB-E4. Exposure should be amortized, but at something less than 100% (*Report*, pp. 65-68, 83-84). (The amortization estimate also requires assumptions concerning the number of days per year and the number of years an applicator will be occupied applying the herbicide.) Strangely, despite this assumption, all of the amortized estimates which the Board considered when working out its reasonable worst-case exposure scenario were derived by assuming *full* amortization.

ARB-E5. In general (the Board made one significant exception to this), a closed application system should be assumed: closed cab tractor; bulk, closed tank, soil incorporation (*Report*, pp. 81, 83).

ARB-E6. The Board appeared to accept the assumption made by both HPB and Monsanto, that exposure resulting from applicator carelessness should be ignored.

Reliance on these assumptions left the Board with an optimistic worst-case exposure scenario. All of the data on which the scenario depended were provided by Monsanto. The data on human subjects which served as the basis for exposure calculations came from two quite different tests, as discussed earlier. That available when HPB calculated its exposure estimates was from patch tests. Although Monsanto carried out further patch tests in 1984, the company also undertook at that time four biomonitoring studies. These involved taking urine samples from applicators during and following mixing, pouring, and applying two different formulations of alachlor (one of which had never been sold in Canada). A total of 48 applications of alachlor were made at three sites: Ontario (four applications in one study), Indiana (eight applications in one study) and Missouri (36 applications in two studies). In Ontario and Indiana, the applicators wore protective gloves (Monsanto's description of the Indiana study states that the gloves were "elbow-length"); in Missouri (36 of the 48 applications), full protective clothing. For the four Ontario applications, open cab tractors were used; for all of the other applications, closed cab tractors were used. The Ontario and Indiana studies used open-pour tank-fill of small containers. One of the Missouri studies used open-pour tank-fill. The other used a closed system tank-fill. In both of the Missouri studies there was soil incorporation of the herbicide.

When it came time to calculate its own exposure levels, the Board thus had two sets of data—from the patch and biomonitoring tests—neither of which formed a firm scientific case. The patch data showed much higher readings than the biomonitoring data but had to be accompanied by a value reflecting the rate of absorption through the human skin into the blood stream, for which there were no data. Although Monsanto had calculated such a value from tests on rhesus monkeys, its applicability for humans was unknown. No known transfer ratio existed. Further, the fact that the highest absorption reading in the rhesus monkey tests had been omitted (ostensibly because too much pressure had been used when rubbing alachlor into the monkey's skin) put the company's procedures into question. The biomonitoring data were potentially more accurate but only when supported by studies which traced the breakdown and metabolism of the product in the specific system in question, in this case the human body. Such metabolism studies did not exist. The Board was well aware that the data themselves were also questionable because, as pointed out during the Hearings by the environmental groups, test subjects were much more highly protected than the normal farmer would be, and the sample size was small.

The Board thus had data on which a wide range of exposure estimates could be based. But unfortunately the evidence available did not provide strong reasons for preferring any one of these data sets over the others. The lowest estimates came from the biomonitoring data, the highest from the patch data assuming 100% absorption. In addition, an intermediate set of estimates could be generated from the patch data assuming the 25% absorption figure which the Board derived from the studies of rhesus monkeys. In the end, the Board chose the biomonitoring data as the basis for its calculations. It felt that the much lower absorption readings provided by these tests were corroborated by the lower dermal absorption and carbon tracing data for the rhesus monkey, and that these data provided the most reliable basis for estimates of actual exposure.[28]

This reasoning involves an apparent contradiction, however. The Board explicitly defended its use of the monkey metabolism data to corroborate the human biomonitoring data: "There are no data to indicate that human absorption and excretion patterns would be significantly different" (*Report*, p. 83). And yet it had previously declared that the metabolism data were unreliable for these purposes: "Although some major differences were found in the metabolism of alachlor in the three species [rat, monkey and human], from 25 to 50% of the administered compound was unaccounted for, or unknown in all the studies. . . . At this time, none of these data can be related to the question of the possible health hazards of alachlor in humans" (*Report*, p. 75).

This may be an oversight, but in the face of problems with both patch and biomonitoring data bases, the selection of the data base which provided the lowest exposure readings and therefore the lowest calculated risk was not strictly scientifically determined. This was rather one of a series of value-based decisions which countered what the Board felt to be the overly conservative assumptions made by HPB in generating its own "worst case scenario." Most importantly, given the fact that all these uncertainties plagued the choice of exposure and absorption estimates, it is hard to see how the choice the Board made could have been said to be reasonable *worst* case. In the end there seemed to be no difference for the Board between "the most reliable" estimate and the reasonable worst-case estimate. But surely these two are not the same when the uncertainties in the data place the former in the middle of a wide range of possible error.

Based on assumptions drawn from animal data regarding the way the human body metabolizes alachlor, the results of analysis of the urine samples were used to calculate the total body dose of alachlor received in each of the 48 applications. For each of the four studies, a mean total body dose was calculated. Evidently, the two Missouri studies were combined and a mean for the two studies was calculated. The mean for the Ontario study was at the top of the range; the mean for the two Missouri studies, the bottom of the range. Average daily doses were calculated from the top and bottom of the range values. This calculation gave an exposure range from 0.02 (top of range, Ontario study) to 0.0006 (bottom of range, Missouri studies) mg/kg/day, unamortized.

From the same data, Monsanto also calculated various ranges of *amortized* exposure estimates, employing a variety of assumptions. In one calculation, the mean average daily dose for the Ontario study was amortized, as was the mean average daily dose for the two Missouri studies. The amortization calculation assumed the application of 4 pounds of alachlor/acre, 100 acres/year, and 40 years of application over a 70 year lifespan. This calculation constitutes *full amortization*.

For the top-of-the-range Ontario study, full amortization reduced the estimated exposure from 20.0×10^{-3} to 31.3×10^{-6}. For the bottom of the range, Missouri studies, full amortization reduced the unamortized estimate from 0.6×10^{-3} to 0.9×10^{-6}. These are the critical estimates for the Board's calculated "worst-case exposure scenario." The Board's scenario referred to the top of the range of the amortized estimates and the bottom of the range of the unamortized estimates. When this formula is applied to the calculations cited here, the relevant numbers are 0.6×10^{-3} and 31.3×10^{-6}.[29] Thus, if no other estimates are considered, these two numbers bound the range that defines the worst-case exposure scenario, with the amortized Ontario estimate forming the bottom of the *scenario's* range, and the unamortized Missouri estimate forming the top of the *scenario's* range.

We emphasize "scenario" in this account to call attention to the transposition that has been effected: the Ontario study had yielded the most exposure but, as a result of amortizing the mean average daily dose, it forms the lower bound in the worst-case exposure scenario; the Missouri studies had yielded the least exposure but, as a result of not amortizing the mean average daily dose, it forms the upper bound in that scenario. We shall comment on the significance of this shortly.

Other amortization scenarios were calculated. One assumed a commercial applicator, with full protective clothing and a closed application system, working for 15 days per year; another assumed commercial applicators, with protective clothing, but open application systems. Since these estimates do not vary significantly from those cited two paragraphs above (*Report*, p. 83), in the interest of keeping the main points in view we shall ignore them here. Taking account of all of the estimates before it, amortized and unamortized, the Board found the worst-case scenario formula (top of the range of the amortized estimates, bottom of the range of unamortized estimates) to yield an exposure range from, approximately, .001 to .0001 mg/kg/day. The figures that we are focusing on, .0006 to .000031, are in the same range and similarly are three to four orders of magnitude lower than the lowest dose at which a tumour was observed in one of the rat feeding studies.[30] (See Figure 1.)

Since citation of this range forms the heart of the Board's estimation of alachlor's risk, our central thesis, that risk estimation is value relative, is best tested by considering some of the judgements that were made (1) in deciding what to look for in determining what a reasonable worst-case scenario is and (2) in arriving at the numbers that form the range, the upper and lower bounds. If the conventional view of risk estimation as being neutral and value-free were correct, all of the critical steps in defining the range would be justified by appeal to empirical data. But it is evident from the foregoing account of the Board's reasoning that much more was going on than measuring and calculating.

In its attempt to assess the risks of alachlor, the Board was confronted with uncertainties at nearly every step along the way. But because the Board was practising regulatory science, it did not have the luxury sometimes enjoyed by laboratory scientists of responding to the uncertainties by refusing to reach a decision—by simply saying, in response to a large question looming before it, that in view of the uncertainties, "We don't know!" It was charged with responsibility for giving an answer. Should the government's decision to cancel alachlor's registration be upheld? Or should Monsanto's appeal against that decision be supported?

An implication of these facts—that it had to reach a definitive answer but, because of the uncertainties, the "facts" did not speak loudly and clearly to indicate what that answer must be—was that it had no alternative but to give answers that served to protect what it thought to be the salient values in the case. This is the first respect in

Figure 1
Alachlor Exposure Estimates

Mg/kg of body wt/day

2.7	HPB patch test, 100% absorption, no protective clothing, no amortization
2.5	LOWEST DOSE AT WHICH A TUMOUR WAS OBSERVED IN RAT STUDIES
0.68	HPB patch test, 25% absorption, no protective clothing, no amortization
0.26	Biomonitoring, no protective clothing, no amortization
0.21	HPB patch test, 100% absorption, protective clothing, no amortization
0.063	HPB patch test, 100% absorption, no protective clothing, full amortization, 15 days exposure/year
0.05	HPB patch test, 25% absorption, protective clothing, no amortization
0.02	Biomonitoring, protective clothing, no amortization
0.0078	Biomonitoring, no protective clothing, no amortization
0.0056	HPB patch test, 100% absorption, protective clothing, full amortization, 1 day exposure/year
0.0047	Biomonitoring, commercial applicators, protective clothing, 15 days exposure/year
0.0042	HPB patch test, 100% absorption, no protective clothing, full amortization, 1 day exposure/year
0.001	UPPER LIMIT OF WORST-CASE EXPOSURE SCENARIO
0.0006	Biomonitoring, protective clothing, no amortization
0.00038	HPB patch test, 100% absorption, protective clothing, full amortization, 1 day exposure/year
0.0001	LOWER LIMIT OF WORST-CASE EXPOSURE SCENARIO
0.000014	Biomonitoring, commercial applicators, protective clothing, full amortization, 15 days exposure/year
0.000031	Biomonitoring, protective clothing, full amortization, 1 day exposure/year
0.000016	Biomonitoring, commercial applicators, protective clothing, full amortization, 1 day exposure/year
0.0000009	Biomonitoring, protective clothing, full amortization, 1 day exposure/year

which the Board's determination of a reasonable worst-case exposure scenario was "value relative." It exemplifies what we referred to in Chapter II as the presence in the risk estimation of *conditionally normative* issues.

The second respect is rather more direct. To develop a reasonable worst-case exposure scenario, the Board had to answer a number of questions which were decisive in the sense that how they were

answered would significantly affect the estimate. Among these decisive questions were many that were simply normative in *kind*. As we
shall see below, these were questions about what is *reasonable to
expect* and *fair to assume*. In the terminology adopted in Chapter II,
the reference here is to questions that are *inherently normative*.

The difference between these two respects in which the Board's
estimate was value relative is that one might hope to put off the need
to invoke values when making decisions in a context of uncertainty
by adopting a delaying tactic. The idea would be to put off the decision until more research has been done, in the hope of reducing or
possibly eliminating the uncertainty. Since uncertainty will always
be with the regulatory scientist, and the need to issue definitive
answers within *some* time constraints will always be there, the delaying tactic is but a stopgap measure. Moreover, as suggested in Chapter II, further research sometimes compounds rather than reduces
uncertainty. Nevertheless, within limits uncertainty can be reduced
and with that the role of conditionally normative questions can be
reduced as well.

By contrast, the second source of value relativity, that which arises
from the need for the risk estimate to answer inherently normative,
and therefore non-empirical, questions, is such that further (scientific) research will be of little help. Here the idea of reducing the
uncertainty has no relevance, except insofar as answering the inherently normative question requires settling a matter of fact. (To say
that such questions are "non-empirical" is merely to say that the
answers do not come from the "facts" alone, not that facts are irrelevant to the answers.)

Conditionally Normative Issues

We shall consider these two sources of value relativity in turn, beginning with that which reflects presence of conditionally normative
issues in the risk estimation. In Chapter II we listed ten issues the
alachlor risk assessors confronted which, because of uncertainty that
surrounded them, should be regarded as conditionally normative.
These issues have been discussed in the preceding pages. Confronted by the uncertainty, Monsanto assessors resolved each issue
in a way that had the effect of minimizing the risk. HPB assessors
followed the opposite course. The Monsanto assessors, one might
say, were optimistic. Since there was no solid scientific basis for a
resolution of any of these issues, the Monsanto assessors resolved
them in a way that contributed to a finding that alachlor is safe.
HPB's assessors were, by contrast, pessimistic or conservative. They
resolved the same issues in a way that contributed to a finding that

alachlor poses a serious health risk. This difference between Monsanto and HPB correlates, of course, with their different institutional positions. Monsanto's interests lay with alachlor's being pronounced safe; HPB's institutional interest is given by its name: health protection. For the most part but with significant exceptions, the Review Board sided with Monsanto on these issues.

Far and away the most decisive issue was that concerning amortization. The difference between a fully amortized and an unamortized exposure estimate is roughly three orders of magnitude. For example, the lowest estimate in Figure 1 assumes biomonitoring, protective clothing, 1 day exposure per year, and full amortization. Based on these assumptions, the calculated exposure estimate is .9 x 10^{-6}. But if this exposure estimate is recalculated, using the same raw exposure data and the same assumptions, with the sole exception that full amortization is replaced by no amortization, then the resulting calculated exposure estimate is .6 x 10^{-3}. Possibly one can reasonably decide on general principles that neither full nor no amortization is appropriate. It follows that alachlor exposure should be amortized at a rate somewhere between 0 and 100%. But given the present state of our knowledge there is no scientific basis for deciding what the exact rate should be within that wide range.

For another example, consider the following question. Should the Board have placed as much reliance as it did on the biomonitoring data submitted by Monsanto? These studies had been completed while alachlor was still under review by Agriculture Canada and HPB, but Monsanto did not release them until the Board began its Hearings, and then released them to the Board. Then there is the fact that the Missouri studies, which were done last, gave by far the most favourable results from Monsanto's point of view. Were they designed in the "state of the art" way that they were in order to generate data that would soften the impact of the relatively poor Ontario results? What of the significance of the fact that the applicators in the studies were all Monsanto employees (some sufficiently high level employees to have appeared as Monsanto witnesses at the Hearings) who were fitted out with protective clothing selected by Monsanto? To what degree can exceptional precautions in applying alachlor and in selection of protective clothing reduce exposure? Monsanto's own witnesses indicated that anywhere from 80 to 95% of the alachlor absorbed was picked up by the hands. Recall the elbow-length gloves.

Obviously, these questions point to a large element of uncertainty in the data base upon which the Board based its worst-case exposure scenario. Was the Board naive not to question whether it should place complete confidence in that particular data base? Whatever the answer, it is clear that in order to defend that confidence, or to defend any other view concerning the reliability of the biomonitoring studies, a position would need to be taken on many issues that are not straightforwardly empirical or statistical, but that instead invoke judgements of value.

To mention the most significant of these, it would be unreasonable to determine how much confidence to place in those studies without reflecting on what is at stake. Suppose one thinks that the price of acting on them, by endorsing a lower estimate of alachlor's risks than HPB arrived at using patch tests, is to create health risks of some consequence that one would very much like to avoid. To the extent one thinks this it would be reasonable to cast a cynical eye on the studies. If, instead, one is primarily concerned with the economic benefits for Canadian farmers and with Monsanto's freedom to make a profit selling alachlor, a different view of the dangers of "over-estimating the risks" will and should take hold. Note that we are considering what is reasonable to do in a context of uncertainty, when the facts do not speak loudly and clearly.

The principle we are depending on here is one the Board employed in its reaction to what it saw as HPB's *over*-estimation of alachlor's risks. The claim then was that in the face of uncertainty HPB had consistently made conservative assumptions and thereby arrived at a conclusion that penalized Monsanto and Canadian farmers—the former by denying its right to market a product; the latter, by potentially withholding from them a means of remaining competitive. This can mean only that in the face of uncertainty it is reasonable to make one's way by having an eye out for the benefits (or, more generally, the values) that are in the balance.

Inherently Normative Issues

In Chapter II we illustrated the idea of inherently normative issues by citing five normative questions confronted by alachlor risk assessors. All five bore on the Board's determination of a reasonable worst-case exposure scenario. The first four raised the issue of fairness. The fifth concerned the identification of a reasonable *worst* case. The fairness issues referred to protective clothing, application systems, the significance to be attached to carelessness on the part of applicators, and the significance of poorly constructed wells.

We have seen how sensitive an exposure estimate is to assumptions made concerning these fairness issues, especially to the assumption made concerning protective clothing, given the finding that 80-95% of applicator exposure is at the hands. The question put by the Board was not the empirical one, whether applicators *do* wear protective clothing, or whether on average applicators do so. It was whether it is reasonable for the purpose of deriving an exposure esti- mate to base the calculation on an assumption that applicators will wear protective clothing. Similar remarks apply to the other "fair- ness" issues. The issues are not resolved by referring to how applica- tors are *likely* to behave or how they *typically* behave. Rather, to decide the issues the Board had to determine whether it would be fair to Monsanto to develop a risk estimation founded on exposure estimates that included exposures incurred owing to failure of appli- cators to wear protective clothing, to exercise due care, and to use closed cabs, and to failure of farmers to ensure that their wells are adequate.

There is, of course, an empirical dimension to the protective cloth- ing issue. The cost and availability of adequate protective clothing bear on the issue. And if in the farming community it were generally the case that pesticide applicators *do* exercise due care, that would be relevant to a decision concerning what is reasonable to expect. But these empirical considerations are inputs to a decision that is inher- ently normative. The conclusion aimed at concerns what for regula- tory purposes *is* fair, not what farmers or others *think* is fair, and not whether people's actions accord with what is regarded as fair.

HPB determined that 80% of applicators wore inadequate protec- tive clothing and was led by this fact to base its exposure scenario on the assumption that inadequate clothing would be worn. But the 80% figure alone would not warrant rejection of Monsanto's and the Review Board's fairness argument. Presumably Monsanto and the Review Board agreed with the figure, or at least did not dispute it. The fact that 80% of applicators wear inadequate protective clothing warrants refusing to discount exposure that results from wearing such clothing only on the assumption, made by HPB, that human health is the main value in the balance. Monsanto's and the Review Board's opposed view, that exposure which results from wearing inadequate protective clothing should not be counted (despite HPB's evidence that adequate protective clothing is seldom worn), depends on their sensitivity to Monsanto's and the herbicide users' economic interests.

The issue of the adequacy of well construction deserves special consideration because it highlights an important aspect of all four fairness issues. If wells contaminated by alachlor are found to have been poorly constructed, or perhaps poorly sited, should that be con- sidered relevant in an estimation of alachlor's risks? Obviously, counting the number of such wells will not resolve the issue. Again, a judgement of fairness and of reasonable expectations is required.

The Board chose to discount the risks posed by the presence of alachlor (and also metolachlor) in the wells and even municipal water systems located in the agricultural areas where these chemicals are used. Studies submitted to the Board indicated that some of the tested wells showed levels of contamination with alachlor and metolachlor that were close to the high end of the range of the Board's exposure estimates for applicators (e.g., 3×10^{-3}), but because they did not exceed this range the Board considered them also to be comfortably within the "margin of safety." Yet a factor in the Board's decision not to treat the well water problem as a serious one was that the problem appeared to be present only in poorly constructed or poorly sited wells, and even then it could be solved with charcoal filtration systems.

An underlying, if not explicitly stated, implication is that, whatever risk might be presented to users of well water contaminated with alachlor, it is to some degree an avoidable and *voluntarily assumed* risk. As such it is not a risk that rates as high in the scale of regulatory concern as might other types of *imposed* risks. This is a major reason for endorsing the Board's view that the issue was not one of *how many* wells were contaminated or at *what levels* they were contaminated. Rather, the issue was whether, because the contamination was preventable by those most likely to be affected, the regulator should take notice of it. That this entered into the Board's thinking is suggested by the fact that in its list of recommendations (*Report*, p. 16) the Board dealt with this matter by suggesting that among the labelling requirements for alachlor there should be a warning "against use in close proximity to shallow wells and surface waters." The assumption here is that the problem is one of providing the information necessary for "informed consent" to risk.

The decision of the Board to exclude from the range of reasonable worst-case exposures to alachlor those resulting from such practices as the non-use of adequate protective clothing, noncompliance with handling procedures specified on the label, and even accidental spills (which were known to occur, and even happened in Monsanto's own controlled study), seems to have been motivated in part by this consideration of voluntariness. As indicated above, whether there is empirical evidence that these things *do* happen was not the Board's concern. Rather, because their happening reflects the voluntary risk-taking of those involved, they are not considered to be a part of the regulator's primary concern. This view is again indicated by the fact that the Board chose to include in its list of recommendations one which specified that a "safety education program" be implemented by Monsanto at its own expense, and that Monsanto provide a free pair of rubber gloves to every purchaser of alachlor.

The assumption underlying the recommendation is that exposure which exceeds the range defined by the Board as reasonable worst-case can be excluded from regulatory concern by ensuring that such exposure is voluntarily accepted.

It is not our view that this assumption by the Board is in any way an erroneous or unsupportable one. Indeed, we believe that one of the significant factors in the weighting of risks, often ignored in standard risk management equations, is the voluntariness or involuntariness of that risk. We underscore this move by the Board only to illustrate one more way in which a fundamentally normative presupposition (viz., that a voluntarily accepted risk is of lesser regulatory concern because of the respect governments should exhibit for the freedom of its citizens to take their own risks if they choose) works to influence what on the surface poses as a scientifically neutral estimation of risk.

As these, often rather mundane, issues are cited, it will occur to one that they are familiar aspects of most risk assessments. Noting the inherently normative character of the issues is something like calling attention to the fact that the Emperor is wearing no clothes. Though there for all to see, it is seldom remarked. It is understandable that his subjects studiously avoided commenting on the Emperor's state of undress. But, given the obviousness, and the obvious decisiveness, of these inherently normative issues, one cannot but wonder at the currency of the classical view of risk assessment as being objective and value-free.

To this point we have been pointing to some of the normative assumptions that the Board made when estimating alachlor's risks. On these same matters, HPB often made *different* normative assumptions. The differences in their risk estimations are in considerable measure traceable to these differences in normative assumptions. Perhaps one can fault either the Board or HPB for making the particular normative assumptions they made, and for mistakenly assuming that they were engaged in an exercise of "objective" risk estimation. But they cannot be faulted for actually making normative assumptions (assumptions about what is reasonable to assume and fair to expect), since without making some such assumptions they could not have produced a risk estimation of alachlor.

We recognize that for many our position concerning the centrality of inherently normative issues in risk assessments will seem counter-intuitive. It will be helpful, therefore, to consider at some length the following objection to the position. There is no *need* to construe any of the issues that must be decided when estimating exposure as inherently normative; moreover, since there are advantages to keeping risk estimations scientifically pure, it is a *mistake* to define any of these issues as inherently normative. Thus, the protective clothing issue, one might say, is best understood as an empirical one. What sort of clothing do applicators actually wear? In particular, do they wear adequate gloves or not? The exposure estimate should be based on the answer to this question.

How shall we reply to the objection? Focus on those issues that refer to "exposure conditions": whether protective clothing is worn, the method of applying alachlor, the amount an applicator will apply on each occasion, the number of applications he or she will make each year, etc. The suggestion is that for purposes of an exposure estimate, these issues should be resolved by looking to actual practice: if this is done, then issues concerning exposure conditions will become empirical and not inherently normative.

But when one does look to actual practice, the most striking feature one observes is the lack of uniformity. Some wear very adequate protective clothing, some wear protective clothing of middling adequacy, and some are exceptionally careless in this regard. The same variation in practice is met with respect to the other exposure conditions. Some applicators apply 10 times more alachlor per year than others do. At most, the empirical study of actual practice would permit the analyst to derive a curve which plots the risk relative to these varying exposure conditions.

Perhaps the advocate of scientific purity's best response to this difficulty is to say, When exposure conditions vary, compute the average. In pursuing this approach, the assessor would need to answer such questions as the following: what is the average amount of alachlor an applicator applies per year? What is the average adequacy of the protective clothing worn? Then the exposure estimate should express the exposure of a person who wears protective clothing of average adequacy and who applies the average amount of alachlor per year.

Let us call this the "averaging approach," to distinguish it from what for now we shall call the "best-case" approach taken by Monsanto and the Review Board, and the "worst-case" approach taken by HPB. These represent three different responses to the need to choose a way of normalizing the exposure conditions. The objection holds that the averaging approach has the advantage that by taking it the risk assessor can maintain the desired neutrality and objectivity. Is this the case?

A distinction must be made between the possibility of maintaining objectivity when defending a preference for one or the other of the three approaches, and the possibility of maintaining objectivity when using the preferred approach in estimating exposure. With respect to the latter, there is no problem. If we ignore the values that must enter owing to uncertainty, then the calculation of exposure based on the averaging, worst-case, or best-case approach might well be value-free. That is, once the approach is chosen, no further inherently normative issues remain. If we are thinking of the values involved in actually calculating exposure, then, the averaging approach is no more neutral than the worst-case and best-case approaches.

But what about choice of the averaging approach itself? Can one justify preferring it to the worst-case and best-case approaches while remaining neutral, scientifically pure? We have seen that Monsanto's and the Review Board's best-case approach to the need to normalize exposure conditions was based on their having priorized economic benefits and "fairness"; HPB's worst-case approach rested on its more or less overriding commitment to human health. Monsanto held that to include in the estimate exposure resulting from applicators' careless practices would be *unfair* and would result in loss of economic benefits associated with use of alachlor. HPB estimated that 80% of applicators do not wear adequate protective clothing and used this fact to support its worst-case assumptions—a line of reasoning which rests on the idea that the primary value at issue is the *health* of the applicators, not imagined fairness to the manufacturer. In this way, Monsanto (and the Review Board) and HPB, perhaps without recognizing what they were doing, departed from neutrality and scientific purity in the choice of normalized exposure conditions. And it is difficult to see how one might justify choice of either the best-case or worst-case approach without appealing to values such as these.

Is there something about the averaging approach which privileges it in this regard? That is, can a preference for it be scientifically pure? Evidently not. Of course, one can refuse to give a reason for preferring it. But then so might those who adopt the other approaches. A decision concerning how to normalize the exposure conditions must be made and the decision will be momentous for the risk estimate. If the estimate is to be taken seriously, then the choice among the three approaches must be a principled one. But the issue is such that the principle appealed to cannot but be normative.

In fact, it is possible to give intuitively plausible reasons for preferring the averaging approach, although we will not claim that they are decisive. What is important about them is that they are normative in character. One may say that what recommends basing an exposure estimate on average exposure is that *both* of the other two alternatives enjoy a measure of plausibility. This fact points to a compromise in the direction of that accomplished by the averaging approach.

One wants to say that there is some merit in Monsanto's claim that extreme worst-case exposure should not form the entire basis for the alachlor exposure estimate. Insofar as the health risk resulting from such exposure is voluntarily incurred, it is unfair to Monsanto to include it in the bottom-line assessed risk of alachlor. To do so would be to penalize Monsanto for the willful failings of others. But one may also say that there is some merit in HPB's appeal to worst-case exposure. It is a matter of being realistic. If 80% of applicators have well-established careless habits and if these habits put them at considerable risk, then the interest in being fair to Monsanto should be tempered by recognition that the health of farmers counts for something too. Adopting the averaging approach is a way of recognizing the merits of both the other approaches.

It is of course not important here to decide among the approaches, but only to see that the choice among them is a normative one. It expresses a view concerning what *should* be done which, if it is not arbitrary, must have a basis in a position concerning values, rights, justice, etc. Possibly the choice will in effect be made in advance by the policy under which the risk assessment is undertaken, or the setting in which it is done. That is, the choice needn't be a personal one, expressing a value to which the risk assessor is personally committed. Whatever their personal predilections, HPB risk assessors are committed to health protection by their job description, and this dictates a distinctive way of normalizing the exposure conditions when doing any exposure estimate. Monsanto's risk assessors, by contrast, operated within a setting which made it entirely understandable that they were sensitive to the "fairness" argument and thus wanted to normalize the exposure conditions so that exposure resulting from failure to exercise due care would not be counted. It is not so easy, however, to identify an institutional source for the approach taken by the Review Board. Perhaps this fact forms a strong reason for questioning the decision to appoint only scientists, or at least no "environmental" scientists, to the Board.

We turn now to the fifth of the inherently normative issues, whether the Board's account of a reasonable worst-case exposure scenario really was "worst-case." The point at issue concerns the sorts of considerations that must go into a determination that *any* such scenario is "worst-case." Recall that the scenario consists in a range. The upper and lower bounds of the range were derived by making calculations on means, in the one case the mean for a range of data derived from the Missouri studies, in the other, the mean for the range of data derived from the Ontario study. We shall raise some questions concerning the identification of both of these bounds.

First, the lower bound, the amortized estimate calculated from the

Ontario study results. We have indicated that full amortization was used, this despite the fact that the Board had rejected "the use of full amortization as equally unreasonable" as no amortization (*Report*, p. 83). Amortization consists in multiplying an unamortized estimate by a number which expresses the fraction of the applicator's life he or she is engaged in applying alachlor. Thus, using the Board's assumptions, if the unamortized average daily dose is expressed in mg/kg/day, then the unamortized estimate will be multiplied by a fraction, of which the denominator is the estimated number of days in the applicator's life (here, 70 years, and so, approximately, 25,550 days), and of which the numerator is the estimated number of those days during which the applicator will apply alachlor (here, 40 days). Thus, on these assumptions, to fully amortize an unamortized estimate of average daily dose is to reduce the estimate by approximately three orders of magnitude. (In the example, the unamortized estimated is multiplied by, approximately, .0015.)

If full amortization as well as no amortization are ruled out, it might seem reasonable to look at the range of data between these two extremes. Following the Board's scenario, but taking account of its finding that something less than full amortization was warranted, the lower bound for its worst-case scenario would fall somewhere between .000031 and .02 (*Report*, p. 80). But where in this range? Note how momentous the answer is. If we decided on a number close to the upper limit of this range, in effect, minimal amortization, we would identify the upper bound in our worst-case exposure scenario as being two orders of magnitude higher than the Board's actual upper bound.

In considering the question, How much amortization is appropriate? it is important to keep in mind that one is looking for a reasonable *worst*-case scenario. Some of the other assumptions employed in arriving at the estimate are these: (1) that 400 lbs. of alachlor are applied each year; (2) use of the mean value rather than that for the worst Ontario exposure test result, which gave a value more than twice the mean value; (3) the "normality" of the conditions of the Ontario study, especially use of protective clothing. Of relevance here is the estimate that 80-95% of exposure is to the hands, and the doubt whether farmer/applicators as opposed to commercial applicators would wear gloves or whether the gloves worn would be of adequate quality. These were certainly not "worst-case" assumptions. So, to a large extent, the scenario must be "worst-case" owing to the decision concerning the percentage of amortization. What number, between 1 and .0015, should be chosen?[31]

There are at least three points to consider in answering this question. First, given the lack of knowledge concerning whether animals or humans develop cancer as a result of chronic or "one-shot" exposure to alachlor, the choice of an amortization percentage is not a purely empirical matter (it is conditionally normative, as shown in the previous section). Second, the Board's decision, to multiply by .0015, yields something closer to a "best-case" than a worst-case estimate, and is only explicable on the assumption that the Board was unusually sensitive to what it saw as the economic benefits and rather less sensitive to the health issue. Third, any reasonable debate between HPB and the Board on this matter would have to focus, not on the sorts of issue that the Board was more or less exclusively occupied with, but on the value issue of whether the risk *estimator* (as opposed to the risk manager) should primarily have in view public health or economic benefits and the freedom of manufacturers to market their products.

Similar remarks are in order concerning the Board's determination of the upper bound. Rather than repeat the point already made, however, we shall stress a special aspect of the Board's choice of the upper bound for its worst-case scenario. We noted earlier that there was a curious transposition: as a result of the decision to amortize the Ontario exposure estimate, the worst case in the Monsanto studies became the lower bound, and therefore, in a sense, "best-case" in the exposure scenario. Why was the decision not made to amortize the Missouri exposure estimate, and report the Ontario results as an unamortized estimate?

The rationale for such a decision might have been that the Ontario results gave more cause for concern than the Missouri results. In the interest of having a *worst*-case scenario, the argument might have gone, it would not be appropriate to manipulate those results in a way that obscured the danger signals they sent out. If it had been decided to leave the Ontario results unamortized, the top of the range for the worst-case exposure scenario would have been .02—again, two orders of magnitude higher than the Board's actual estimated upper bound.

As noted, the Board decided to identify as the worst-case exposure scenario the bottom of the range of unamortized estimates and the top of the range of amortized estimates. Many of the estimates in the former range were provided by HPB and were based on patch tests. The amortized estimates were calculations derived from Monsanto's four biomonitoring studies. The Board reasoned that since some amortization was appropriate, and all of the amortized estimates fell below the unamortized estimates, the worst-case exposure scenario should specify a range of values that extend no higher than the lowest unamortized estimate. Since the estimate calculated from the mean daily dose value for the Missouri studies (.0006) was roughly at the bottom of the range of unamortized estimates, it was picked as the top of the range for the worst-case exposure scenario. But, as we have indicated, keeping the range of amortized estimates below the lowest unamortized estimate depended on the Board's decision to use calculations based on full amortization, despite its conclusion that this was not appropriate.

The issue is whether to leave unamortized the Missouri or the Ontario data. Again, our argument is not that the Board made the wrong choice. We only call attention to the sort of choice it is—that it is not an empirical or mathematical matter but a matter of values—and to the extreme sensitivity of the risk estimation to the way the choice is made.

We shall conclude this long discussion of inherently and conditionally normative issues by considering the cumulative effect on exposure estimates of differing views concerning these issues. The estimates shown in Figure 1 range from 2.7 to .0000009 mg/kg/day. It is only a slight simplification, which does not distort the overall picture, to say that these widely divergent estimates represent calculations made on the same raw exposure data.[32] If one takes those data, and assumes a patch test should be used to measure the exposure, that there was 100% absorption of alachlor deposited on the patch, that an applicator would wear no protective clothing, and that exposure should not be amortized, then the calculated exposure estimate becomes 2.7 mg/kg/day. If, instead, one begins with the same data, but assumes that biomonitoring should be used to measure exposure, that full protective clothing would be worn, that exposure should be fully amortized, and that applicators are exposed only 1 day per year, then the calculated exposure estimate becomes .0000009 mg/kg/day. What we have found is that all of the assumptions responsible for the six orders of magnitude difference between these two exposure estimates are either inherently or conditionally normative. Patch test versus biomonitoring, rate of absorption, and full versus no amortization, are conditionally normative issues. The protective clothing issue and the issue concerning the number of days per year the applicator will be exposed are both inherently normative. In other words, the choice between concluding that an applicator's exposure is 2.7 or that it is .0000009 mg/kg/day is not one that the alachlor risk assessors did or could have made by appealing to the facts alone. Making the choice involved and required invoking values, in particular, estimates of the relative importance of health and economic benefits. The significance of this conclusion for the classical view of risk assessment is obvious. Anyone who thought that applicator exposure would be 2.7 mg/kg/day would regard alachlor as a dangerous product; anyone who thought the exposure would be .0000009 mg/kg/day would regard the product as remarkably safe.

Consideration of an Objection

We have argued that uncertainties lead to conditionally normative conclusions in risk assessments because the assessor has no alternative but to presuppose values when there is inadequate factual basis for resolving an issue. Suppose one objects to this position in the following way: risk assessors can avoid these conditionally normative issues by building the uncertainty into their conclusions. One way to do this is to cite a range rather than a specific number. For example, it is uncertain whether patch tests or biomonitoring yields the more reliable measurement of exposure. But if a patch test indicates that the exposure is .005 mg/kg/day, and biomonitoring indicates that the exposure is .0005 mg/kg/day, then the scientifically defensible (and value-free) conclusion is that the exposure falls in a range between .005 and .0005.

The same objection may be made to the parallel claim regarding inherently normative issues. Different value assumptions (health versus market fairness, say) dictate different ways of normalizing exposure conditions: concern for health supports an assumption that applicators will not wear adequate protective clothing; concern for market fairness supports an assumption that they will do as advised on the product's label. Suppose that on the former assumption the exposure estimate is .001 mg/kg/day, whereas on the latter assumption (everything else equal) the estimate is .0001 mg/kg/day. The most scientific of risk assessors will work with the range given by the two values, .001 and .0001. Recognizing that choice of a value within the range could not be made on purely scientific grounds, the assessor would refuse to make the choice.

In considering this objection it must be noted that the strategy of reporting ranges rather than exact numbers will work only if applied to all of the uncertainties and all of the inherently normative issues that plague the risk assessment. It is no good to say, simply, that other things being equal, the exposure is somewhere between .005 and .0005, depending on what one thinks about the merits of patch tests; and that, other things being equal, the exposure is somewhere between .001 and .0001, depending on what one wants to assume about protective clothing. The ranges need to be amalgamated. Thus, in the alachlor case the risk assessor needs an *exposure* estimate. So the range which, according to the objection, the assessor should cite will be marked at one end by the estimate that results from making the most conservative assumptions; at the other end, the estimate that results from making the most optimistic assumptions. We have indicated that the two ends of the range are 2.7 mg/kg/day and .0000009 mg/kg/day.

Now the claim is that the citation of this range constitutes a purely scientific assessment of the risk of alachlor that avoids normative judgements of both the conditional and inherent sorts we have alleged to be determinative in the alachlor controversy. Anything that goes beyond the mere citation of the range is not a scientific assessment of risk, but is already a crossing of the threshold to the second, normative stage of risk evaluation (i.e., the question of the acceptability of risk). But this is precisely the claim we wish to challenge. We cite two important reasons why we find this view inadequate.

(1) The position is that if we strip the values from the science then we can conclude no more than that alachlor exposure will fall somewhere in the range between 2.7 and .0000009 mg/kg/day. The same limitation applies to the conclusions to be drawn from the animal feeding studies. All we can say is that statistically (or "biologically") significant tumours (of various types and of uncertain significance with respect to carcinogenicity) were observed in rats in a range of increasing incidence, beginning at 2.5 mg/kg/day. These two ranges, however, as important and as scientifically reliable as they may be, do not constitute in any sense whatever an assessment of risk to human beings. Scientifically speaking (on this account) the most we can say is that human beings *might* be exposed to alachlor at levels comparable to those that produced tumours in rats, or the highest exposure *might* be many orders of magnitude lower. Does this constitute a high risk? A low risk? Or no risk at all? We simply have no answer to this question. To move toward an answer, that is, toward an assessment of *risk*, we need to fill in the whole series of gaps in the data—the metabolism questions, the amortization questions, the patch test/biomonitoring questions. In the absence of this data conditionally normative judgements have to be made before we can actually have an assessment of risk.

Even if we knew that humans metabolize alachlor exactly like rats, and we had all the amortization and absorption questions answered, we still would have no purely scientific basis for saying what risks, if any, they faced in the application of alachlor. This is because the question of protective clothing worn by the applicators would remain and this, as we have pointed out, simply is not a scientific question. It may be scientific to the extent that the range of actual behaviours is empirically determinable. But we still must decide which of the behaviours within the range we are going to choose as the base for estimating the risk. Risk is always relative to the conditions of exposure. So, apart from a specification of the possible exposure conditions we think important, there can be no risk assessment. We have to decide whether to include the risk faced by a person who does not wear adequate protective clothing and this decision is a decision based on the normative consideration, among others, of the fairness of doing so. Until this normative decision is made, there simply is no risk assessment.

We do not question whether science, stripped of all value judgements, can tell us *something*. The question, rather, is whether it can provide an estimation of *risk*. When we confine ourselves to what value-free science can conclude concerning alachlor, we aren't left with anything very useful—at least not useful to the regulator, who must decide whether alachlor poses a risk to human applicators, and *which* applicators. The question the regulator faces is whether the fact that there is a slight overlap in the two ranges established by the feeding and the exposure studies provides any reason for concluding that there is a risk to human applicators of alachlor. The science itself simply does not answer this question. In order to answer this question a decision *must* be made concerning all those gaps in the data and, most importantly, concerning which segment of the exposure range to take most seriously. These decisions depend upon the normative judgements we have identified. If the promise of a purely scientific risk assessment is that it can replace completely the imputed emotionalism and "subjectivity" that underlies public perception of risk by something more substantial and reliable, it simply is unable to deliver on that promise.

One may say that uncertainties and inherently normative issues loomed unusually large in the alachlor case. That remains to be seen. But in any risk estimation there will be uncertainty, and in any risk estimation that centres on an exposure estimate there almost certainly will be inherently normative issues. So the same principles will be involved as in the alachlor case. The determinedly "scientific" assessor will be pushed to citing a range rather than a number. The only possibility is that in other cases the range may be somewhat narrower. Equally, in other cases the range may be even broader.

(2) The second reply to the objection raises a deeper issue. The objection assumes that "risk" is an objective fact determinable at least ideally by pure science. Where this ideal cannot be reached because of uncertainties, it proposes that risk assessments acknowledge the uncertainties by citing ranges within which the objective risk resides. The idea is that the risk falls somewhere within the range, although one cannot say where. But an implication of the presence of inherently normative issues in risk estimations is that the

"risk" of a product is not an objective fact. That is, it is not a property of the product itself, or of the way the product is used.

The distinction we have in mind is fundamental. Alachlor, for example, has a chemical structure of some kind. Even if, owing to limitations of knowledge, we are unable to say precisely what that structure is, nevertheless we can be confident that it has *some* structure and that this structure is an objective property of alachlor itself. But people also have reactions to alachlor: some like it, some dislike it. Alachlor's being liked or disliked is not an objective property of it; the "liking," one may say, qualifies the person who likes it, not the substance liked. Of course, alachlor has the property of *being liked*. But this property is either subjective or relative; remove the individual who likes alachlor and the property of being liked is removed as well.

The "risk" of alachlor is in this sense either subjective or relative, and not objective. This is most easily seen if a few simplifying assumptions are made. Suppose that there were no uncertainties in the alachlor exposure estimate and that the only inherently normative issue was that concerning protective clothing. If the exposure conditions are normalized by assuming that no protective clothing will be worn, the exposure estimate is .5 mg/kg/day; if they are normalized by assuming that adequate protective clothing will be worn, the estimate is .005 mg/kg/day. Under these suppositions and assumptions, the objection we are considering recommends that the risk estimation be based on an exposure estimate that consists in the range, .5 to .005 mg/kg/day. If "actual" exposure falls at the top end of this range, alachlor is very risky. If it falls at the bottom end, alachlor is considerably less risky.

Focus on the top end of the range, .5 mg/kg/day, the number that supports a judgement of high risk. The "high risk" attributed to alachlor under these conditions is an objective property of alachlor only in case the exposure estimate, .5 mg/kg/day, refers to *actual* exposure, the actual amount of alachlor to which real applicators will be exposed. But this exposure estimate results from normalizing the exposure conditions. It does not refer to the exposure of an actual applicator of alachlor, but to what the exposure would be if an applicator were not to wear protective clothing. And the claim that exposure should be estimated under an assumption that applicators will not wear protective clothing is a normative claim.

True, under our assumptions, applicators who do not wear protective clothing and thus subject themselves to .5 mg/kg/day of alachlor, put themselves at high risk. (And *this* risk is an objective fact.) But when risk assessors issue a verdict of high risk, they are going beyond the hypothetical claim concerning the exposure that will result from wearing no protective clothing. The reference, rather, is to normalized exposure. The high risk to which applicators who refuse to wear protective clothing subject themselves can be identified as "the risk of alachlor" only in case .5 mg/kg/day is the right exposure estimate. And, as we have seen, whether it is the right estimate depends on how an inherently normative issue should be settled.

What underlies normalizing the exposure conditions by assuming that no protective clothing will be worn is the high value one places on human health and the relatively low value placed on the allegation of unfairness to Monsanto of including the case of the careless applicator in the range of alachlor's risks. Monsanto and the Review Board reversed this order of value priorities and thus excluded the condition of no protective clothing. So the estimate that alachlor is extremely risky has built into it a value judgement concerning the importance of human health and fairness. This is not to say that *if* health is the right value to invoke, then the objective risk of alachlor is very high. It is to say that claiming that the risk of alachlor is very high *incorporates* a claim that health is the right value to invoke, and that as a result the risk of alachlor is not an objective property of alachlor but is instead relative to that value judgement.

To understand this point it is helpful to clarify the concept of "risk" itself as it is used in risk assessment. It should be clear that risk is not something that inheres in things themselves, but rather in a relationship between things. One imposes a risk upon another. Defined technically, a "risk" is a certain probability (X) of one state of affairs or object (A) causing a harm to another state of affairs or object (B) *under a specifiable condition (Z)*, multiplied by the magnitude of that harm (Y). If the conditions (Z) under which A and B relate to each other change, then, obviously, the variables X and/or Y may change, and so the overall risk in the situation changes. This is just an analytical way of pointing out that risk is always relative to *exposure*.

It is clear, then, that when one inquires about the "risk" of something like the herbicide alachlor, this is merely a shorthand way of asking what harm it is likely to cause to something else (e.g., rats, human applicators, or users of farm well water) when it is used in a certain way (e.g., in certain concentrations, with certain frequencies, using certain clothing). We do not know what the risk is until we specify these conditions. If we had perfect information about the properties of alachlor and its effects upon the human organism under various conditions, it might be said that we know on a purely scentific basis the objective risk inherent *in each of these conditions*, without appeal to any normative assumptions.

In most cases where risk assessment is called upon for help, however, particularly in the regulatory situation where we need to know
the risk of something like alachlor, we need to know more than the
"objective" risks of alachlor under various conditions. We need to
know which of the conditions should be taken as representative for
the purposes of stating what the risk of alachlor is—which the regulator either should or should not be concerned about. This is one of
the inherently normative issues that bedevilled the alachlor debate:
should the case of the applicator who refuses to wear adequate protective clothing be taken as the normative exposure condition which
states the risk of alachlor?

This point sometimes appears in risk debates in a different guise.
Monsanto representatives at the Hearings as well as some of the
Review Board members made statements reflecting their belief that if
some applicators were careless in their handling of alachlor, or if
some users of contaminated well water had defective wells, then the
risk was in the behaviour, or in the well, not in the product alachlor.
Is this a spurious claim? Not at all. This was simply their way of
making the argument that these conditions ought not to be taken as
normative. HPB officials and the environmentalists took the opposing normative view that these conditions reflected the risk of
alachlor itself.

If one thinks value judgements are "subjective," in the sense that,
like tastes, there is no disputing them, then the claim that the risk of
alachlor is very high (or very low) should be seen as "subjective"
too. If one thinks that views concerning the relative importance of
health and economic benefits are amenable to rational discussion,
and that there are defensible and indefensible views on this matter,
then the claim that the risk of alachlor is very high should be seen as
"relative" (to the value judgements on which it depends) but not
"subjective." But in either case, the effect of the dependence of the
risk estimate on positions taken concerning inherently normative
issues is that the risk estimate itself is inherently normative.

We have been assuming that on the assumption that protective
clothing is not worn the correct exposure estimate is .5 mg/kg/day,
whereas on the assumption that protective clothing is worn the correct estimate is .005 mg/kg/day. Can risk assessors succeed in their
pursuit of a value-free risk estimation by refusing to decide the protective clothing issue, and content themselves with citation of the
range, between .5 and .005 mg/kg/day?

In one sense yes, but in another sense no. Yes, in the sense that the
assessor has helped further the risk assessment by establishing a
range of possible exposure outside of which there is no risk to be
taken seriously. If the whole of this range had indicated safe levels of

exposure, a claim of "no risk" would have been substantiated; in this case as well, of course, citing the range would count as a risk assessment. But such was not the case with alachlor. Since the high end of the exposure range crossed over the lower end of the dose-response range in animals a decision had to be made whether this constituted a risk of alachlor. Assessors who, aspiring to be value-free, refuse to take this decision are in effect refusing to answer the question which motivated the entire enterprise: is alachlor risky? In this sense they are not offering a risk assessment at all.

The Determination of a Reasonable Margin of Safety

The Board found a three to four orders of magnitude gap between the bounds of its worst-case exposure scenario and the lowest dose at which a tumour was observed in one of the rat studies. The Board's decision that this represents a reasonable margin of safety was expressed in one sentence, and no argument was given to support it. The Board must have thought the conclusion self-evident. But this self-evidence is seen against a background of assumptions. In part, the gap gives comfort because of the uncertainties in the estimates at both ends of the gap—the applicator-exposure estimates and the lowest estimated dose at which a tumour was observed in the rat studies. (In light of the uncertainties, one feels the need for a wide margin for error.) But how much comfort one should find is a function of one's view of those uncertainties, on one hand, and, on the other, of the "costs" incurred if it turns out that one has charted the wrong course through the uncertainties. One can readily imagine circumstances in which a three to four orders of magnitude gap would feel not comfortable but chilling.

Two final points. First, the Board elsewhere had questioned the reliability of data concerning the carcinogenicity of a chemical in animals for inferring its carcinogenicity in humans. For the Board, the relevance of this view was that any inference that a particular chemical presents as much risk to humans as it does to animals is without foundation. That is, the Board focused on the possibility that the chemical is *less* dangerous to humans. But it might as well have focused on the other possibility, that the chemical is *more* dangerous to humans. When the Board confidently identified the margin of safety as three to four orders of magnitude, it implicitly assumed, contrary to its earlier disavowal, that extrapolation from the animal to the human case is reliable, and ignored the possibility that alachlor is more dangerous to humans than to rats. Consideration of this point introduces an additional uncertainty to the calculation of the margin of safety.

Second, those exposed to alachlor are also exposed to other carcinogenic substances. It is possible, even probable, that some of these interact with alachlor in the sense that the carcinogenicity of alachlor is increased as a result of the exposed individuals also being exposed to those other substances. The implications of this fact, which is uppermost in the minds of those who worry about chemical farming, were not explored by the Board. This is a worrisome aspect of all risk assessments which are closely tied to the isolated environment of the research lab in which the real-world background of cumulative risk is excluded.

We have shown that there is some reason to question the Board's decision that the gap presented by the margin of safety is three to four orders of magnitude. So it is not just an abstract speculation that the gap might actually be less than the Board supposed. If made at all, the decision that the gap affords a reasonable margin of safety should be made in full awareness that there are defensible perspectives from which the gap is seen to be much less than three to four orders of magnitude. In these circumstances, it would be the height of folly to pronounce the margin of safety adequate without taking full account of the human costs of being wrong.

Given a slightly different set of what, in our view, are perfectly plausible value assumptions, the Board's worst-case exposure scenario would be seen as closer to best-case than worst-case. Had an equally reasonable (and given HPB's values, a more reasonable) worst-case scenario been developed, the margin of safety provided between the exposure values the scenario identifies and the lowest dose at which a tumour was observed in one of the rat studies should have been found inadequate. Thus, it is our view that the Review Board did not demonstrate the reliable *scientific* basis it claimed for its conclusion that the Government's estimation of alachlor's risks was excessive. Consequently, its conclusion remains scientifically unsubstantiated. The Board merely demonstrated the implications of the value framework it chose to embrace.

Contrary to the Board's own claim that its proceedings were "primarily scientific in nature" (*Report*, p. 27), we have shown that a great deal more was involved in their deliberations and in these issues in general, than science. What is at stake in this debate, as in all risk debates, are conflicts among fundamental social and moral values, which cannot be resolved by scientific inquiry alone.

V

The Role of Values in Choice of a Risk-Benefit Standard

So far we have concerned ourselves with the risk assessment stage of the alachlor debate. We have argued that the classical model's assumption that risk assessment is a value-free enterprise does not accurately reflect the assessment of alachlor's risks and that the nature of that enterprise is such that it cannot be value-free, even in "ideal" circumstances. We noted earlier that the classical model envisages a second and distinct stage of risk management, in which the *acceptability* of the assessed risk is determined.

On the surface, however, the Review Board's conclusions about the risks of alachlor obviated the need to move on to the risk management stage. It concluded that the risks of cancer from alachlor were negligible, if present at all, given the margin of safety in expected levels of exposure. Given that there was little or no risk, there was no need to determine whether the level of risk was acceptable and no need to invoke any normative standard of acceptable risk. Nevertheless, the Board attempted to strengthen its case by arguing that the government had erred in not taking into account the benefits of permitting farmers to use alachlor, and in not weighing these benefits against what the government perceived as the risks. It proceeded to an inquiry into the appropriate standard to use, decided that risk-benefit was that standard, criticized the government for not using the risk-benefit standard, and actually applied the standard to reinforce the conclusion that alachlor's registration should be reinstated.

The Board's reasoning in arriving at the conclusion that a risk-benefit standard is the appropriate one to use in determining whether alachlor is acceptably safe is instructive. It reveals something about the Board's underlying values, assumptions, and methodology. And, through this, it reveals something about the theory and practice of Risk-Benefit Analysis itself.

From the perspective of the classical model, the Board's invocation of the risk-benefit standard reflects a move from the risk assessment stage to the risk management stage—to the question of whether the level of risk "discovered" by the objective inquiry of the first stage is "acceptable." However, what our analysis of the risk estimation of alachlor in the preceding chapters demonstrates is that the Board's adoption of the risk-benefit approach to the risk management question fed back into and significantly influenced that estimate. We have seen that critical decisions at various junctures in the risk assessment, such as the estimates of worst-case exposures, were motivated by the fear of the *costs* (i.e., loss of benefits) that might accrue from an over-estimation of alachlor's risks. One of the important implications of this is that it calls into serious question the classical view's assumption that risk assessment (including risk estimation) and risk management form two distinct and independent stages. It demonstrates that the values to which one will appeal at the latter stage are likely to feed back into the assessment at the former stage.

To put the matter bluntly, the Review Board invoked the risk-benefit standard in a double-barrelled fashion. First it used the prospect of the potential benefits to be lost from an over-estimation of alachlor's risks to *lower* the estimation of those risks. Then, having significantly discounted the magnitude of the risk, it moved on to the risk management stage where it placed them into the balance once more against the benefits of alachlor use. In effect, the Review Board was counting the benefits *twice*—once in order to reduce the weight of the risks, and then once again when balancing those benefits against the diminished risks. We suspect that this "double counting" of benefits against risks may be commonplace in risk assessment and risk management contexts dominated by the risk-benefit standard.

In considering the question of standards for acceptable risk, the Board drew a distinction between "absolute" and "relative" standards. In effect, it equated the idea of a relative standard with that of a risk-benefit standard: a risk is acceptable or unacceptable only in relation to the risks and benefits associated with alternative courses of action. It is acceptable if it is outweighed by the expected benefits of taking the risk. So, the most acceptable risk (i.e., the "safest" course of action) is the one in which the net balance of benefits over the risks is greatest.

The Board understood an absolute standard to be one that identifies a particular level of risk as unacceptable in itself, quite apart from the potential benefits to be expected from taking the risk, and apart from the risks of alternative actions. An absolute standard identifies a cutoff point: risk at or above that level is deemed unacceptable. Thus, with respect to alachlor, one might hold that the risk it poses to applicators and mixers is unacceptable if the rat feeding studies disclose that the worst-case exposure of alachlor to applicators and mixers is at or above the level at which one of the rats developed a tumour. The "absolute" standard at work here would be, "No tumours are acceptable, regardless of the benefits." The Review Board held that absolute standards are inevitably arbitrary. That is, it held that wherever the cutoff is placed, no good reason can be given for placing it there rather than above or below that point. For example, suppose that the one-in-a-million standard is adopted: a risk is unacceptable in the event it poses a one-in-a-million chance of some untoward effect. Then no good reason can be given for not having placed the cutoff instead at one-in-a-half-million, or at one-in-ten-million.

The Review Board's Argument for the Risk-Benefit Standard

The Review Board concluded that the appropriate standard to use is a relative one, in particular, the risk-benefit standard. Its reasons for this conclusion can only be inferred, but there are at least three possibilities.

(1) *Self-evidence.* There is reason to think that the Board regarded as obvious on the face of it that a risk is acceptable if it is outweighed by compensating benefits, and therefore that to determine acceptability of risk the appropriate procedure is to compute the risks and benefits and then consider whether the latter outweigh the former. It is a natural extension of this view of acceptable risk to hold that for any set of risks that we have to choose among, the one with the highest benefit-risk ratio is the most acceptable.

(2) *The arbitrariness of the alternative.* There is also reason to think that the Board saw the issue in the following way. There are just two candidates, an absolute and a relative standard. Absolute standards are unacceptable because they are arbitrary. This leaves only a relative standard. So, since the Board equated the idea of a relative standard with that of a risk-benefit standard, the risk-benefit standard was chosen by default. That is, since the alternative to the risk-benefit standard is inevitably arbitrary and therefore untenable, the risk-benefit standard is the only one worthy of consideration.

(3) *The "objectivity" of the risk-benefit standard.* Members of the Board, with the exception of its Chairman, were scientists. In addition, one was a risk analyst. It seems clear that an attraction of the risk-benefit standard from their point of view was what they perceived as its objectivity.

There are two dimensions to this line of argument. First, the values to which one appeals in applying the standard appear to be objective; there seems to be little occasion for dispute concerning their status as values. As was seen above, the risks considered were those of humans beings getting cancer owing to exposure to alachlor. That this constitutes an untoward outcome seems beyond dispute and, as it were, as much a *fact* as the fact that alachlor is manufactured by Monsanto. The benefits to which they appealed were, as will be noted below, economic in nature, primarily yield and price effects. Here as well we seem to be dealing with hard facts: that higher yields and lower production costs constitute benefits seems beyond dispute.

Second, and related to this, application of the standard also seems to be an objective matter. Estimation of the risks was evidently seen by the Board to be a scientific inquiry. Similarly, estimation of the price effects of removing alachlor from the market appears to call for nothing more than fairly elementary economic analysis, involving most especially consideration of the effect on the price of metolachlor if Ciba-Geigy has a virtual monopoly. In a similar way, estimation of the yield effects of removing alachlor from the market seems to rely on nothing more than scientific experimentation involving trial plots on which corn and soybeans are grown, with some of the land treated with alachlor, some treated with metolachlor, and some untreated. If the *alternatives* to a risk-benefit standard involve appeal to values that are considerably more contentious, and determination of the impact of cancellation of alachlor on those values is less subject to scientific control, then it is understandable that Board members, given their scientific background and predilections, would reject those alternatives in favour of the risk-benefit standard.

It is evident that these three lines of reasoning in support of the risk-benefit standard are complementary.

How did those who saw the issue of standards in a different way arrive at their view? This primarily refers to the position of the government but also to Ciba-Geigy, the environmentalist groups, and Mrs. Van Engelen. Although there was some vacillation on the part of government spokespersons, none of these individuals and groups agreed with the Review Board on the issue of standards for acceptable risk.

Mrs. Van Engelen did not think that benefits to Canadian farmers should be regarded as compensating her for the risks posed by presence of alachlor in her well water. She supposed that she had a right to non-contaminated water which should not be overridden by appeals to economic gains achievable only by allowing agricultural practices that had as a side-effect the contamination of her well water. "I did not choose that ingredient [alachlor] to feed to [my children]," she said to the Review Board. "Why should I have to be exposed to alachlor to drink and to eat when I did not choose to be in that situation?" (*Hearings*, pp. 1614ff.). Evidently, Mrs. Van Engelen believed that the issue of the acceptability of alachlor's risks had to do with her right to choose whether or not to take those risks.

Similarly, the environmentalist groups thought the question of safety involved more than simply the presence or absence of offsetting benefits. If alachlor caused cancer in humans then it should not be marketed, and a lower level of public health should not be bartered for enhanced farm income or for lower food prices. These parties to the debate, as well as the government, often put this point in terms of the claim that "safety only is the issue, not benefits."

As noted, the government vacillated. It first indicated that the decision concerning alachlor should be made without reference to the benefits of alachlor's use. Then it seemed to acknowledge the relevance of Agriculture Canada's (as opposed to Health and Welfare) taking account of benefits. But, overall, government spokespersons were scarcely able to articulate any account of the standard that was used to decide that the risks posed by alachlor were unacceptable. Nor did anyone speaking for the government manage to say much that was helpful concerning the standard that should be used for such purposes.

Nevertheless, government actions strongly suggest that two complementary standards were actually used. One is alluded to for illustrative purposes above. This is the principle that the risk of alachlor is unacceptable if the rat feeding studies disclose that the worst-case exposure of applicators and mixers to alachlor is at or above an amount that corresponds to the minimum dose at which a tumour was observed in the rat feeding studies. In effect, then, the government tacitly employed a NOEL standard.

The NOEL standard is an application of the generic principle known as the "zero risk" standard. The government was saying that if humans are exposed to an amount of alachlor that is sufficient to cause cancer in rats, then there is no margin of safety. But to regard a risk as unacceptable when there is no margin of safety is to invoke the principle that a herbicide is safe only if it presents zero risk. The underlying idea is that with respect to exposure to alachlor there is a threshold below which alachlor will not induce cancer in humans. The NOEL standard would require bringing human exposure to alachlor below that threshold.

The second standard employed by the government is referred to in a letter sent to Monsanto by the Food Directorate of Health and Welfare on November 12, 1982, warning Monsanto that it is likely to recommend to Agriculture Canada that alachlor's registration be cancelled: "It is the policy of the Health Protection Branch to eliminate or reduce to a minimum human exposure to potential carcinogens . . ." (*Report*, p. 35). On the face of it this identifies a goal, rather than a standard. But the goal can without strain be recast as a standard: the risk of alachlor is unacceptable in case there is an alternative to it that would serve to lessen human exposure to potential carcinogens. For convenience we shall refer to this as the Minimum Exposure Standard. Since HPB had concluded that alachlor was a potential carcinogen and that metolachlor was less harmful than alachlor, application of the standard yielded an unambiguous recommendation to cancel registration of alachlor.

The government's assumption here appears to be that herbicides themselves are necessary but that its regulatory task is to ensure that the herbicides used are those that present the lowest risk to humans. Since the only hazard under consideration is cancer, the goal of reducing risk to a minimum translates into that of reducing human exposure to carcinogens to a minimum.

The minimum exposure standard may be seen as an application of the well-known ALARA standard ("As Low As Reasonably Achievable"). This view results if we suppose that in endorsing use of herbicides the government was assuming that it would be unreasonable to pursue reduction of the risks presented by herbicides to the point of an outright ban. What is "reasonably" achievable in the way of risk reduction then is fixed by the result of applying the minimum exposure standard.

If the assumption that herbicides are necessary is removed, then the minimum exposure standard appears instead as a zero risk or NOEL standard. That is, now the goal of reducing human exposure to carcinogens to a minimum is not constrained by a need to accept the minimum level of risk presented by herbicide use. If less risk results from elimination of herbicides, then the unconstrained minimum exposure standard implies that herbicides should be banned.

The NOEL standard is an "absolute" standard in the Alachlor Review Board's sense. The minimum exposure standard is also absolute if the assumption concerning the necessity of herbicide use is not made. In these circumstances, neither contemplates trade-offs for benefits, much less benefits in the sense of financial gain for farmers.

If the minimum exposure standard is constrained by an assumption that herbicide use is necessary, and if the argument for its necessity involves appeal to the benefits of such use, then it functions as a relative standard.

Evidently the Review Board did not recognize that the government had been guided by these standards. Despite the fact that the sentence from the Food Directorate's letter to Monsanto is quoted in the *Report*, the Board takes no account of the standard implicit in the sentence or of the government's use of the NOEL standard. It gives no reasons for rejecting these standards, nor does it explain its preference for a risk-benefit standard. Of course, on the *Board's* estimation of the risks of alachlor the NOEL standard would not support Agriculture Canada's decision that the risks of alachlor were unacceptable. But the issue here is the assumptions that underlay choice of standard for determining acceptable risk, not those that underlay the estimation of risk.

The Review Board's Options

The Board construed the benefits of alachlor in negative terms, as consisting in the avoidance of certain losses resulting from alachlor's removal from the market. The significance of construing benefits in this way is best seen after noting the limited sorts of benefits the Board took account of. The Board considered only economic benefits, in particular, yield effects and price effects. Thus, in calculating the benefits of alachlor use it looked at two things: first, the impact on yields of corn and soybeans of removing alachlor from the market; second, the impact on the cost of growing corn and soybeans of removing alachlor from the market. The Board referred to this second consideration as a "price effect" because it mainly identified the differential cost of production with the difference between the price of alachlor and the price of its principal substitute, metolachlor. If Ciba-Geigy, the manufacturer of metolachlor, had a virtual monopoly, what would be the impact of that on the price farmers had to pay for herbicides?

An effect of construing benefits negatively, as the avoidance of losses associated with removing alachlor from the market, is that benefits become relative to alternatives. For completeness, the Board identified four options that it had to choose among: cancel alachlor, register metolachlor; cancel metolachlor, register alachlor; register both; cancel both. Two of these involve cancellation of alachlor. But the consequences of adopting either would differ significantly from those of adopting the other. As a result, if the benefits of alachlor are identified with the losses entailed by its not being on the market, then it doesn't make any sense to speak of the benefits of alachlor in absolute terms. Rather, there are benefits of alachlor relative to a state of affairs in which metolachlor is available as a substitute for alachlor, and other benefits relative to a state of affairs in which metolachlor is not available as a substitute.

A further point of some importance is that one cannot say *a priori* that these alternative states of affairs will represent losses. Ciba-Geigy might well have argued that despite the monopoly the price of metolachlor would be less than the price at which alachlor would be sold, if it were on the market, and that use of metolachlor produces more corn and soybeans than does use of alachlor. In that case, the claim would be that the "benefits" of alachlor, relative to a state of affairs in which metolachlor is available as a substitute, are negative; that is, prices would be higher and yields lower.

For the most part, when it considered benefits the Board focused on the benefits of alachlor relative to metolachlor, that is, to the state of affairs in which metolachlor is available as a substitute. Its consideration of the benefits of alachlor relative to a state of affairs in which both alachlor and metolachlor are removed from the market was extremely sketchy. It noted that "continued use of [chloracetanilides] appears necessary if current levels of production of corn and soybeans are to be maintained in Canada" (*Report*, p. 108) and concluded that "If alachlor and metolachlor are both removed from the Canadian market, the likely outcome would be serious adverse impacts on corn and soybean production. These sectors generated over $684 million in farm receipts in 1986" (*Report*, p. 108).

But no attempt was made to translate these claims into dollar estimates of benefits. Among the estimates needed are the percentage reduction of corn and soybean production that would result from removing alachlor and metolachlor from the market, and the impact on production costs of not using these herbicides. If one followed this exercise through, would one find that the benefits of alachlor relative to a state of affairs in which both alachlor and metolachlor are removed from the market (that is, avoiding the losses associated with removing them) clearly outweigh the risks of keeping these herbicides on the market?

The Board held that the ultimate arbiter of acceptable risk is the Canadian public and, given its endorsement of the risk-benefit standard, interpreted this to mean that the crucial question is "whether or not Canadian society would find the balance [of risks and benefits] acceptable or unacceptable" (*Report*, pp. 24-25). But it is not obvious (nor could it have seemed obvious to the Board) what Canadian society would actually conclude if it balanced the risks and benefits of removing both alachlor and metolachlor from the market and compared the result with the balance of risks and benefits associated with marketing both of these herbicides.

We shall consider below the implications of the Board's conclusion that it is for Canadian society to decide whether the balance of risks and benefits associated with a herbicide is acceptable. Here we only note that by adopting this view the Board effectively rejected the risk-benefit standard as the ultimate arbiter of acceptable risk. Since no constraints are placed on how the public is to decide the issue, the effective standard is not whether the benefits outweigh the risks but, simply, social acceptance: a risk is acceptable if the public find it so. Arguably, the most qualified spokesperson for the public in this regard is the Minister of Agriculture, who had already pronounced alachlor unsafe. This suggests that at the point when the Board decided that the public is the ultimate arbiter it either should have changed the course of its inquiry, to investigate whether the Minister had violated recognized procedures for a regulatory agency in a democratic society (a political and legal question which the largely scientific Alachlor Review Board was not competent to answer), or it should have concluded its deliberations and informed Monsanto that its appeal was rejected.

Thus, the question to be faced is, what motivated the Board to jump to the conclusion that the option of removing both alachlor and metolachlor from the market was not tenable. The answer appears to be that the Board was committed to the status quo in agriculture. That is, present agricultural practice involves extensive reliance on chemicals, both to fertilize the soil and to keep down pests and weeds. Although there is a vocal and enthusiastic minority who advocate alternative approaches—organic or ecological agriculture— the dominant agricultural technology is founded on chemicals. Removal of both alachlor and metolachlor from the market would probably make sense only to one who had reason to reject reliance on this dominant technology. Our review of the Board's deliberations strongly suggests that it was unprepared to question the dominant technology in any serious way.

Critical Questions Concerning the Review Board's Reasoning

A critical study of the Board's decision requires answering a number of questions that are raised by the foregoing account of its manner of arriving at and applying the risk-benefit standard. These questions are listed below and in each case we supply what we take to be the answer. In some cases the answer is fairly obvious. In other cases a certain amount of argument is necessary to explain our view. The questions are of two kinds. The first set, which we address in this section, refer to the Board's reasons for endorsing a risk-benefit standard. The second set, which we address below, refer to the reasoning by which it sought to apply that standard. Despite the fact, noted earlier, that the Board's decision was not actually based on weighing risks and benefits, it structured its report as if this were what it was doing.

(1) Is the disjunction on which the Board settled between absolute and relative standards exhaustive? That is, are all standards for determining acceptable risk of one or the other of these two sorts? Our answer is that it depends upon how these terms are defined. If an absolute standard is defined as one that is necessarily arbitrary, or necessarily bad, and a relative standard is defined as a cost-benefit standard, then, as we point out below, the disjunction excludes other plausible alternatives.

But even if the disjunction were exhaustive it does not follow that the decision-maker must employ either an absolute or a relative "standard" in order to determine acceptable risk. "Standards" are a specific kind of reference point for decision-making. One who decides what to do by reflecting that it would be "cowardly" to do otherwise, or "ignoble," is not deciding by referring to a standard of "cowardliness-avoidance." Or, at least, to claim that this is a "standard" would be to stretch that notion far beyond its normal use. Also, a decision based on appeal to past practice is not founded on reference to a standard, as that term is normally understood. And, finally, not all *goals* are readily translated into the language of standards, except in a trivial sense.

The practical relevance of this observation concerning the restrictiveness of the idea of a "standard" is met when a further question is put: need the standard to which one appeals in determining acceptable risk be the *sole* normative consideration employed in reaching a decision? It seems obvious that it needn't be. Rather, the standard might be appealed to in order to establish a *prima facie* view of whether the risk presented by some product or activity is acceptable. Then the final decision whether it is acceptable might be made by bringing in other considerations. This bounded use of standards better fits ordinary notions of reasonableness in decision-making than does the catechetical use involved in supposing that the result of applying a standard must dictate one's decision. An example of this bounded use of standards is suggested in our answer to the second question below.

One may feel disturbed by this reference to appeal to other considerations, thinking that, in a regulatory context, those who seek to have a product registered have a right to know by what standard their product will be judged. But the manufacturers' right in this regard is simply a right to know the ground rules, including the sorts of consideration that will be deemed relevant in the decision whether to register the product. That is, they have a right to know how the decision will be reached. But this can be provided by clarifying what the "other considerations" will be and how they will influence the decision and does not require identification of a standard that will be the sole determinant of the decision.

(2) Is an absolute standard necessarily arbitrary, that is, of such a sort that no good reason can be given for preferring that standard to some other? In arriving at the conclusion that a risk-benefit standard is the appropriate one to use, the Board assumed that absolute standards are inevitably arbitrary. At best, however, this is a highly contentious position. The Board reached its view of the matter by considering a special case of absolute standard, one which identifies a level of acceptable risk in quantitative terms. It does seem reasonable to say, in effect, with the Board, that if the standard is pegged at one-in-a-million, then no good reason can be given for preferring that to a slightly higher or slightly lower ratio. But against this two observations are decisive:

(a) Not all standards are of this quantitative sort. Both of the standards mentioned above as having been actually employed by the government, the NOEL and minimum exposure standards, are absolute standards. In a way, each associates the level of acceptable risk with a number. But in each case the number that identifies that level is derived from an independent consideration.

With the NOEL standard, the level is set experimentally, by determining the minimum dosage at which rats develop a tumour. Whatever that number is found to be, if the question is asked,"Why not pick a slightly higher or slightly lower number?" a clear answer can be given: that, rather than some other number, was picked because it identifies the minimum dosage at which rats develop cancer. This contrasts sharply with a "one-in-a-million" standard, in that with this type no principle is employed to fix the ratio. It is the absence of a principle that makes the charge of arbitrariness plausible.

Of course, one might quarrel with the principle embedded in the NOEL standard. Why regard the minimum dose at which rats develop a tumour as salient? But even if in the end one finds reason to reject this standard it cannot be rejected simply as being arbitrary. For there is *some* evident sense in invoking the embedded principle. In part the relevant explanation would be that reasonable assumptions about the way rats and humans metabolize alachlor indicate that we have cause for worry if human exposure to alachlor is likely to reach the level at which rats begin to develop cancer. At this point a debate might ensue—some claiming the standard should be more demanding or conservative, since there are cumulative effects (possibly humans are exposed to other carcinogens that contribute to the same sort of cancers that alachlor does, and in calculating exposure levels we need to consider all such carcinogens), others claiming that the standard should be more relaxed or liberal, since there is reason to think that rats metabolize alachlor much more slowly and less completely than humans. But this debate doesn't place the issue in the realm of the arbitrary: it is nothing like defending preference for "one-in-a-million" over "one-in-999,999."

With the minimum exposure standard the situation is somewhat similar. The non-arbitrariness of the standard is even more transparent. Up to a point, the question, "Why adopt the standard that the risk posed by a herbicide is acceptable if the herbicide would result in less human exposure to carcinogens than the alternatives?" answers itself. The regulator does not need to justify aiming to reduce human exposure to carcinogens. So far, again, the situation is entirely unlike that presented by a "one-in-a-million" standard and the charge of arbitrariness is inappropriate. If exposure to carcinogens is not an "evil," and reduction of such exposure not a good, then the entire regulatory exercise falls outside the sphere of reason and no distinction between reasonable and unreasonable decisions can be made.

A new issue is introduced, however, by consideration of this standard. It is defined by making clear what really underlies worry about the minimum exposure standard. One suspects that exposure to carcinogens is practically speaking inevitable, and that in general risk is an inescapable fact of life, so that to adopt the standard (or others like it that focus on reducing risks to a minimum) would lead to paralysis.

Our first observation concerning this issue is that it does not bear on the objection that the minimum exposure standard is arbitrary. It points in a different direction by calling into question the reasonableness of invoking a standard that apparently would force rejection of products and activities which, while presenting some risk, however slight, promise significant benefits. A reply to this objection is needed.

But even before the reply is considered one can see that the objection assumes, first, that in principle *any* risk is acceptable, provided only that there are compensating benefits outweighing those risks, and, second, that there are conceivable benefits sufficient to justify taking any risk. That is, it assumes the ultimacy of the risk-benefit standard. But if the argument for that standard uses as one of its premises the arbitrariness of all standards other than risk-benefit, then it would be begging the question to invoke the assumption of the ultimacy of the risk-benefit standard to establish that some particular absolute standard is arbitrary.

We offer now a reply to the objection that the minimum exposure standard would have the unfortunate result of requiring us to forgo significant benefits when a condition of achieving those benefits is assumption of some risk, however slight. We note the point made above, that adoption of an absolute standard need not involve a commitment to let the result of applying the standard dictate the regulatory decision, so that no considerations other than those tied to application of the standard can be taken into account.

Noting the reasonableness of taking a more relaxed view of standards is not an indirect way of reintroducing the risk-benefit standard. (Instead of saying that the decision should result from balancing the risks and the benefits, one seems to be saying that one should invoke an absolute standard—which appeals to a level of risk that marks the point where the acceptable becomes unacceptable—and then be prepared to decide other than as the standard recommends in case there are sufficient offsetting benefits.) This for three reasons:

1. Some standards *will* identify risks that are so serious, or aspects of actions that are so important to avoid, that for all practical purposes *any* benefit (or any benefit of a certain sort, for example, monetary benefit) would be insufficient to justify taking the risk.

2. Even when the risk is not of this extreme sort the "other considerations" appealed to when making a decision that overrides the standard may not refer to "benefits" but may be of a completely different sort. Examples of such "non-benefit" considerations are given in our answer to the first question above.

3. Even when the "other considerations" appealed to are, properly speaking, "benefits," the weight of the standard that identifies unacceptable risk might be such that promised benefits would be sufficient to override the standard only in case they were very substantial.

In other words, the weight of the standard might be such that in referring to benefits one is not balancing risks against benefits but checking to see whether, contrary to expectation, there are benefits associated with accepting the risk of such magnitude as to defeat a presumption that, because the standard for acceptable risk is not met the risk should not be taken. The important difference between this case and the risk-benefit standard is that when a risk-benefit standard is employed there is no such presumption: risks go on one side of the scale, benefits on the other, and the task set is to see or judge which side goes down. By contrast, here, application of the absolute standard makes a *prima facie* case that the risk should not be taken.

Then the appeal to benefits has a special burden of proof if the *prima facie* case is to be overturned.

The minimum-exposure standard seems to be of this third sort. Certainly one can envisage in the abstract benefits of such magnitude as to make it reasonable to prefer a risk that fails to satisfy the standard. But because this would be the case only in very special circumstances (where the benefits would be considerable and adhering to the standard would yield only a slight reduction in exposure to carcinogens), within the normal range the standard is not relative but absolute.

(b) But even when an absolute standard is of the sort envisaged by the Board—"one-in-a-million," that is, where the cutoff level is not fixed by appeal to a principle or other objective consideration—nevertheless it need not be arbitrary. There are two cases to consider, one of which is relevant to the issue at hand, whether absolute standards are necessarily arbitrary, the other of which leads to the answer to question 3 below.

The first case is the normal one. The absolute standard seems arbitrary because no reason can be given for preferring the cutoff identified by the standard to one slightly above or below it (why not "one-in-999,999"?), but that arbitrariness only occurs within a middle range. At either end of the range it seems reasonable to hold that there is no arbitrariness. That is, although no reason can be given for preferring a one-in-a-million to a one-in-999,999 chance, if the choice were between one-in-two and one-in-a-million it would be clear. We do not need persuading that a 50/50 chance of cancer is too high, and that it is much more reasonable to opt for one-in-a-million as the acceptable risk cutoff.

The implication is that absolute standards of this sort are arbitrary when compared with certain alternatives and not at all arbitrary when compared with certain others. Nor does it matter whether one can cite a reason for preferring a one-in-a-million to a fifty/fifty standard. Even without access to a reason one will be more certain of the preferability of the former standard than one will be of any number of issues that present themselves for settlement in the course of doing a risk assessment or when attempting to decide whether an assessed risk is acceptable.

(3) Are arbitrary standards necessarily bad standards? That is, suppose there is no good answer to the question, Why peg the level of acceptable risk at the point where it has been pegged? Why not put it at a slightly higher or slightly lower level? Does it follow that the standard should be rejected?

To see that arbitrariness is not necessarily bad, consider speed limits. Suppose that there is no reason for preferring that the speed limit for driving on a certain type of road be set at 100 rather than 101 or 99 km/hr, or even 105 or 95, so long at least as whatever the number picked there is some uniformity throughout the Province. One should not infer from this supposition that it is a mistake to set the limit at 100 km/hr. The important underlying consideration is that even if there is no reason for preferring one of these numbers to any of the others, there *is* good reason for having a speed limit. So, one reasons, the fact that it would be just as reasonable to pick a different limit works both ways: it is just as reasonable to pick 100 as it is to pick some other number close to that one. And, of course, there are good reasons for *not* picking numbers that range too far from 100—speed limits of 200 or 30 are unreasonable on the expressway.

Similarly, if one defines acceptable risk so that a one-in-a-million chance of some untoward outcome is acceptable, then the fact that the ratio is, within some range, arbitrary does not imply that it should be rejected. Again, the fact it is arbitrary works both ways: there is no reason for picking it, but there is no reason for picking any other. That fact, coupled with the premise that *some* ratio must be picked, forms an argument for the ratio. To be sure, the argument does not establish the preferability of the ratio over one slightly more or less stringent, but it does make the case for having some absolute standard within the "arbitrary" range. Consequently, one cannot appeal to the arbitrariness to justify preference for a relative standard, as the Board did.

Of course, if the Board had an independent reason for preferring relative and rejecting absolute standards, then it would have a reason for being unhappy with arbitrary standards. But the Board's strategy was to attempt to justify that preference by *assuming* that arbitrary standards are necessarily bad ones. (Absolute standards are arbitrary, arbitrary standards are bad ones, the only alternatives to absolute standards are relative standards, therefore, relative standards are preferable.) The foregoing considerations present a counter-example to that assumption.

The Review Board's Consequentialism

We want now to draw a few morals, suggested by the responses to the foregoing three questions. We have considered the Board's reasons for endorsing a risk-benefit standard and for rejecting any absolute standard. The discussion has indicated that absolute standards need not be arbitrary and that arbitrary standards need not be bad standards. These conclusions have been reached by taking a wider view of the value considerations relevant to decision-making than that taken by the Board.

Thus, we took seriously Mrs. Van Engelen's claim that she had a right to uncontaminated well water and that, within reasonable bounds, this right is not subject to being overridden by the financial benefits farmers realize by using herbicides that also contaminate well water. Our assumption was not that she definitely has this right, but only that her assertion of it cannot be ruled out of court as unjustifiable or irrelevant, as it appears to have been in the Alachlor Review Board Hearings.

More generally, we have suggested, in effect, that the Board's view of absolute standards as necessarily arbitrary depends on an assumption that the only considerations relevant to practical decision-making are those referring to the consequences or outcomes of an action or policy. The "wider view of value considerations" alluded to above is a view that encompasses aspects of actions or policies other than their consequences or outcomes. Mrs. Van Engelen's assumed right to uncontaminated well water is an example of such "extra-consequentialist" considerations. Other examples include the following:

1. Appeals to loyalty or other commitments to individuals and groups that result from past relationships.

2. Appeals to distributive justice, that is, to the need to achieve a desired *distribution* of advantages and disadvantages, such as risks in the community, or to secure an adequate *minimum level* of well-being for all members of the community.

3. Appeals to virtues or character traits, to the need to ensure that individuals and groups act so as to express the sorts of character that they, on reflection, admire or ought to admire. Here the suggestion is that when individuals or groups inquire into the sorts of character that they do or should admire, they might decide to abstract from the consequences of being one or another sort of person (or group) and simply confront the issue as one of self-definition, to be resolved by a reflection that turns inward, to what they are and aspire to be, rather than outward, to the *results* of being or doing one thing or another.

This "reflection on self" is available to an entire society as well as to separate individuals. To some extent the recent Canadian debate on free trade with the United States was a masked exercise in such reflection. The free trade debate illustrates the point that *what* an individual or group finds salient in the way of consequences (and thus, in a risk estimation context, its identification of hazards) is often an outcome of self reflection into how the individual or group wishes to define or identify itself. We shall note below how such reflection is involved in the position of those who take seriously the Board's fourth option, that of banning both alachlor and metolachlor and of moving away from chemical farming.

On one hand, the Board needed to take a narrow, consequentialist, view of the value considerations relevant to practical decision-making in order to reach the conclusion that absolute standards are necessarily arbitrary, and thus to convince itself of the appropriateness of the risk-benefit standard. On the other hand, once it adopted that standard it had, as it were, a filter through which none but consequentialist considerations could pass. For to judge an action or policy by reference to its risks and benefits is to judge it by reference to outcomes, by reference to what will happen in consequence of taking the action or adopting the policy. "Benefits" is the name we give to a certain sort of "good consequence"; "risks," the name we give to a certain sort of "bad consequence." A result is that considerations of the sort cited in the above list do not influence the decision.

Why did the Board take this narrow view of the value considerations relevant to practical decision-making? The beginnings of an answer are suggested in our reference to the backgrounds of Board members. All, except the Board's Chairman, were scientists. It seems evident that they interpreted their task so that they could bring their scientific expertise to bear. This involved their seeking to position themselves as neutral outsiders, who bring no values of their own to their deliberations.

Our suggestion is that this view of themselves and of their role on the Board (a view which dictated an attempt to maintain a stance of "objectivity") motivated the narrowing that we are referring to, the focusing on consequences, and in particular on "risks" and "benefits." It made the Board less responsive to rights claims, such as that made by Mrs. Van Engelen, or to appeals to "the kind of society we want to be"—founded on self-reflection—that lead to proposals for less dependence on chemical farming.

The attempt to maintain a stance of neutrality and objectivity points one toward what look like the "factual" aspects of the alachlor issue. From this perspective, appeals to rights and to "the sort of society we want to be" are appeals of the wrong sort. They point to values, concerning which there may be disagreement, and to aspects of the issue that are difficult, if not impossible, to measure. By contrast, as we have noted above, "risks" and "benefits" appear to be factual aspects of the issue, non-controversial and, generally, measurable.

Behind this stance of neutrality and "objectivity" lurks a distinctive understanding of rationality, an assumption that adopting the risk-benefit perspective, or something very like it, which focuses on the apparently factual and measurable, is a condition of being rational. An implication, not spelled out by the Board, is that those who take a wider view of the value considerations relevant to the alachlor issue are irrational.

Our suggestion is not that the Board's underlying assumption concerning the conditions for rationality was wrong, but only that it was decisive. It led to the Board's rejection of absolute standards and dictated its endorsement of the relative, risk-benefit standard. And thereby it ruled out of court the considerations of a non-consequentialist sort invoked by at least three parties to the debate: (1) The government and the environmentalist groups, through their tacit advocacy of one or another absolute standard; (2) Mrs. Van Engelen, through her appeal to a right of third parties or the general public to uncontaminated well water; and (3) those (unrepresented) groups who embed opposition to alachlor in a global objection to chemical farming, an opposition which is itself defended by appeal to the sort of society we should want to form (the way we should want to relate with one another, technology, and nature).

The Review Board's Value Framework: Rationality

This suggestion affords a fresh view of the alachlor debate that supplements and reinforces our remarks concerning risk assessment in preceding chapters. It focuses attention on the hidden debate, an underlying disagreement that was not recognized and therefore not discussed. This disagreement concerns the meaning of rationality. For if we say that the Board operated with one view of rationality, then we must recognize that the other individuals and groups alluded to above, just because they appealed to principles the Board could not accept, operated with a different view of rationality. The idea is not that these others held a narrow view of their own that identified different values and depended on a different methodology which rendered references to risks and benefits irrelevant. It is that they held a wider view, that encompassed the sorts of considerations to which the Board was sensitive, but went beyond those to invoke extra-consequentialist principles of the sort mentioned above.

Confirmation of our suggestion is found in a consideration of the Board's way of *applying* the risk-benefit standard. Here we ask two questions:

(1) In applying the risk-benefit standard to determine whether the risks of alachlor are acceptable, why did the Board define risks and benefits so narrowly, so that virtually the only risks considered were those to humans of getting cancer from exposure to alachlor; virtually the only benefits, those associated with yield and price effects? That is, what explains this attenuated view of the values at issue?

If one were asked to produce a list of the hazards that might be associated with use of a herbicide, it would not be difficult to fill a page. Possible carcinogenic effects would be mentioned, but so would toxic effects, dangers associated with the activity of mixing and applying the herbicide, possible pollution of streams and rivers, health effects borne by others than mixers and applicators, including animals as well as other humans, consequences for the soil, and so on. Here two points should be noted: (i) probably none of these other effects are as readily quantifiable as are carcinogenic effects. (ii) Also, although arguably all represent "evils" to be avoided, and hence have the negative character essential to anything that is to count as a hazard, the evil of cancer is dramatically less contentious and, therefore, more "fact-like" than any of the others.

A result is that if one feels constrained by one's conception of rationality to adjudicate the alachlor issue from the stance of a neutral and value-free observer who attends only to the facts, then one will find comfort in a conception of the "risk" side of the risk-benefit equation which restricts the risks under consideration to those associated with carcinogenic effects suffered by alachlor mixers and applicators.

A similar point may be made by looking to the benefit side of the equation. We have noted that virtually the only benefits considered were yield and price effects. But again here the list of possible benefits is nearly endless. Why then focus on just these two? Yield and price effects *appear* to be readily quantifiable and indisputably "benefits," and hence preserve what some regard as the illusion of engaging in a scientific, neutral, and value-free inquiry.

We stress the word "appear" to open space for asking whether price and yield effects should actually be counted as benefits. The relevance of the question is that the Board's not having entertained the possibility further confirms the hypothesis that its inquiry was guided by a view of rationality not shared by all of the groups whose views were at odds with the Board's. The price and yield effects of alachlor were regarded by the Board as benefits of alachlor use. These "benefits" primarily accrue to the farmers who use alachlor; they are, in this sense, direct benefits. The risks associated with alachlor use are imposed by these same farmers. Insofar as these risks are borne by others than those who impose them and, in addition, are *involuntary* risks, those others are required to carry a burden in order that the farmers might gain. But in these circumstances the price and yield effects enjoyed by the farmers as a result of alachlor use might well be viewed by others as *ill-gotten gains* and when so viewed are not properly classified as "benefits" in risk-benefit calculations. It is evident that Mrs. Van Engelen would regard the suggestion that she should submit to contaminated well water, in order that farmers might enjoy the price and yield effects of alachlor, in this light. Even though a "farm-wife," she departed from that role in the Hearings, to the extent of apparently pitting herself against her own farmer husband, who refused to get involved in her protest.

But the important point here is not that the price and yield effects should be regarded as ill-gotten gains. It is rather more general. Reference to the possibility that the so-called "benefits" are ill-gotten gains illustrates the complexity of the idea of a benefit. Benefits are not facts. An outcome that benefits an individual or group should not for that reason alone be treated as a "benefit" in a risk-benefit calculation. A judgement is called for, by which it is determined either that the gain to that individual or group is one to which it is *entitled*, or at least that the gain is not one to which the individual or group is not entitled. Thus, we do not count the gains that accrue to robbers in calculating whether we should allow them to practise their trade.

The relevance of this reference to the complexity of the idea of the benefits of alachlor is that it is a complexity that would pose a difficulty for one whose conception of rationality was of the sort we are attributing to the Board. A natural response to the difficulty would be to ignore the complexity. And this, as we have seen, is what the Board did.

To summarize: the narrow conception of risks and benefits has two aspects. It consists in restricting one's view of the risks and benefits associated with alachlor use to those that are non-contentious, "fact-like," and quantifiable. And a corollary of the attempt to represent risks and benefits as fact-like is that the indicated complexity in the idea of a benefit (and, for that matter, in the idea of a risk) will be skirted. This narrowing *would* make sense to those whose understanding of rationality leads them to suppose that reason lies with quantification, a disinterested attention to the facts, and avoiding value judgements.

(2) In considering how, once the risks and benefits of alachlor are estimated, the two are to be weighed against one another, why did the Board conclude that *Canadian society* must judge whether the benefits outweigh the risks so that the risks are acceptable?

According to the classical model of Risk-Benefit Analysis, for each alternative all of the risks are to be homogenized and assigned a number, and the benefits are to be quantified in the same way, so that a benefit-risk ratio may be calculated. Then that alternative is to be chosen which bears the highest benefit-risk ratio. On this model, there is no occasion for referral of a risk-benefit issue to "the Canadian society." The risk- and benefit-estimation phases, carried out by experts, supply the numbers. The risk-benefit standard supplies the algorithm. The decision, sometimes relegated to a distinct "risk management" stage of the inquiry, does not call for judgement but for calculation, which can also be done by an expert, or, if this is something different, by a computer.

The Board saw the impossibility of applying the classical model. Too many of the relevant factors escaped quantification. And even if a number could have been assigned to each side of the risk-benefit equation, the number on one side would have been, as it were, a number of apples; on the other, a number of oranges—more specifically, a number of cancers and a number of dollars. Dollars cannot be reduced to cancers, and there is a certain arbitrariness, if not repugnancy, in reducing cancers to dollars. The result is that there is no possibility of applying the algorithm.

Perception of these facts was expressed by the Board in an acknowledgement that specifically *quantitative* Risk-Benefit Analysis was not available as a way of settling the alachlor issue. This led it to refer to its task as *qualitative* Risk-Benefit Analysis. But what can this mean with respect to the balancing of risks against benefits? If the balance of risks and benefits cannot be calculated then it must be judged. How is this judgement to be exercised?

As noted, the Board answered by referring to "Canadian society" as the arbiter. What explains this leap? That is, why did the Board feel constrained to identify Canadian society as the arbiter at this critical juncture? The answer seems evident once an implication of the conception of rationality we are attributing to the Board is made explicit. If being reasonable consists in being neutral and value-free, sticking to the facts, and, so far as possible, quantifying, then the most or only truly rational way to determine acceptable risk is that described by the classical model. The impossibility of applying the model to the alachlor issue implied, then, the impossibility of being rational at the critical juncture when the risks were to be weighed against the benefits. At that critical decision point, reason has no purchase, the experts must be silent, and the choice must be made by "the people."

In other words, an implication of the conception of rationality we are attributing to the Board is that, given the impossibility of applying the classical model, the decision whether the benefits of alachlor outweigh the risks is a subjective one and as such the only appropriate group for making the decision is the Canadian public. We have indicated above, however, that it is misleading to say that the task then set the Canadian public is that of weighing risks against benefits. If they are to judge, then they are to judge, and whether other considerations than those of risks and benefits are to be invoked is also something they will judge. The decision reached, however it is reached, cannot be defended as being *rational*, but only as *democratic*.

But even if the Canadian public, or the Minister of Agriculture acting for the Canadian public, does restrict its attention to risks and benefits, it is still misleading to characterize the decision as a result of applying the risk-benefit standard. For the Canadian public will not be "weighing" risks against benefits, but will simply be deciding *in light of* the risks and benefits. It may find benefits of such magnitude or significance that *any* risks would be worth taking to secure those benefits. In that case, there is no weighing, risks are irrelevant, and benefits alone win the day. Or it may find that the risks are so serious, so unacceptable, that *no* conceivable benefits could outweigh them. Then, too, there is no real "weighing," benefits are irrelevant, and risks alone win the day. In both cases, if there is a recognizable standard brought into play it is an absolute standard. These are, to be sure, but abstract possibilities. But whether those possibilities are actualized is altogether for the designated arbiter, the Canadian public, to say. As we have indicated above, the effective standard is that of social acceptance.

The Review Board's Value Framework: Liberalism

The social acceptance standard, which the Board, somewhat inconsistently, invoked, gets its appeal from the idea of voluntariness. We suppose that in general a self-imposed risk must be acceptable because we wish to respect the autonomy of the individual or group who choose to assume the risk. The underlying idea is the liberal principle, famously defended by John Stuart Mill, that (within limits) society must not seek to coerce any of its members when the sole rationale for such coercion would be the member's own good. The principle applies to the alachlor case: if Canadian society is identified as the ultimate arbiter concerning the safety of alachlor, then, since they are the risk-bearers, the group who are asked to assume the risk are the ones who are authorized to determine whether it is acceptable.

A rather different application of the principle was extensively entertained in the Hearings. It was maintained that for the most part it is the farmers who are put at risk by alachlor, and it is those same farmers who would decide whether to use the herbicide. Thus, if accurate, comprehensible information about the risks posed by alachlor is made available to the farmers, and there are no other coercive circumstances impinging upon their choices, the risks are voluntarily assumed. The question was then put, but not answered: why not reinstate alachlor's registration, regardless of the extent of its carcinogenic effects, and leave it for the corn and soybean farmers to decide whether the risks are acceptable to them?

One answer, of course, is that in fact the farmers are not the only group at risk. Mixers and applicators are often employees of the farmers who decide to put alachlor on their fields. There are others in the community whose well water may be jeopardized. The run-off into streams and rivers affects people at some distance from the fields where alachlor is applied. Animals are affected. And no doubt other remote risks could be mentioned. (In large part, though, this answer was not available to the Board, since they chose to concentrate almost exclusively on the risks borne by applicators and mixers.)

There is a second answer, however, that the Board did not enter-tain. We cite it here because it suggests an amplified account of the underlying assumptions on which the Board acted in considering the alachlor issue. Assume, as the Board concluded, that alachlor gives the farmer who uses it a competitive edge. If, then, it is marketed in a Province and used by some of the corn and soybean farmers, the oth-ers, who are worried about its carcinogenic effects, will be in a bind. If they choose not to use alachlor they will do less well than those who choose to use it and thus will suffer a penalty. The penalty will derive from the fact that alachlor is registered. Whether the threat of a penalty for not using alachlor will be experienced by particular farmers as coercive is impossible to say, without a fuller specification of the circumstances. But it is evident from this consideration that, to the extent alachlor does give the farmers who use it a competitive edge (the extent to which it actually does have benefits), the risks it imposes are not fully voluntary. At the extreme, if the competitive edge afforded by alachlor were so great that a farmer who chose not to use it could not continue in business, then the risks it imposes would scarcely be voluntary at all.

It is evident that this second answer is associated with a certain (sceptical) view of the free market and with a disposition to find coerciveness in situations that appear, on a superficial and formal view, to be voluntary. There is an ideological divide between those who are impressed with answers of this sort and those who are not. It seems plausible to suspect that the Board's silence on the matter, their failure to indicate even awareness of the possibility of answer-ing the question in this way, results from their being situated on the free enterprise side of the divide.

The suggestion that emerges from these reflections is that the Board was operating with a bias in favour of profits and growth, a classically "liberal" mind-set that favoured non-interference with "market forces" and that identified the economic benefits promised by such non-interference as the most significant ones to attend to when

evaluating policies and considering appeals against regulatory decisions.[33] This suggestion gains strong and direct support from the particular way in which the Board defined the benefits to be derived from
use of alachlor. It also gains support from the fact that, as we observed
in Chapter IV, the Review Board's definition of the "worst-case" conditions was motivated by a concern for market fairness to Monsanto.

The Review Board's Value Framework: Pro-Technology

We conclude this chapter with a brief account of the significance of
the Board's manner of confronting the option of banning both
alachlor and metolachlor. As noted above, the Board evidently did
not take this option seriously. It neglected to estimate the real benefits that might accrue from banning both herbicides. Thus it
presented the Minister of Agriculture with no basis on which to
judge whether the risks of this course of action outweigh the benefits, or, rather, whether its net benefits are greater than those attached to the other options. We put the point this way because Chapter 6 of the *Report*, in which the Board presents its conclusions concerning the benefits associated with the use of alachlor, opens with
the following words:

> This chapter reviews the evidence provided to the Board on the
> subject of the benefits to Canadian farmers from the use of
> alachlor. It is in the context of these benefits that the Board
> believes the Minister must judge whether the risks of allowing
> alachlor to be used in Canada are acceptable or unacceptable.
> (*Report*, p. 96)

The Board's failure to estimate the risks and benefits of the fourth
option, banning both alachlor and metolachlor (or, more realistically
described, placing controls on their use), is correlated with its narrow definition of risks and benefits. If the only risks under consideration are carcinogenic effects, and the only benefits those associated
with yield and price effects, then there is not a great deal to be said
about the risks and benefits of this fourth option. First, there are no
obvious risks (that is, carcinogenic effects) associated with it. Second,
the major benefits, including, for example, less contamination of the
soil, are of a sort that the Board did not include in its working definition of risks. (And we have cited above the extra-consequentialist
consideration that was obviated at the start when it was tacitly
decided to restrict attention to "risks" and "benefits.") A result is
that the Board could do little more than point to the value of the
annual corn and soybean crops produced in Canada and suggest

that the size of the crops would be reduced if the fourth option were adopted. There would be "serious adverse impacts on domestic corn and soybean production" (*Report*, p. 108).

One might say, of course, that although the Board did not look closely at the risks and benefits associated with the fourth option, this was rendered unnecessary by its prior determination that alachlor poses no or virtually no risk to humans and that, when compared with the option of marketing metolachlor but not alachlor, retention of alachlor was found to have positive price and yield effects. But note that this defence of the Board's neglect of the fourth option depends on accepting its unnaturally narrow construing of "risk," as consisting only in the likelihood of carcinogenic effects, and on accepting also its restriction of "benefits" to price and yield effects. Those who are worried about the growing reliance on "chemical farming" base their worry on a much broader understanding of risks and benefits.

We have indicated above that in our view the Board's abrupt dismissal of the fourth option suggests an assumption on its part that it would be wrong to tamper with the status quo in agriculture. This means that it entered the inquiry with a bias in favour of chemical farming which it had no desire to question. The strongest evidence we can offer for this view is the fact that nowhere in the 40-plus-volume record of the Hearings nor in the Board's carefully put together *Report* does one find *any* serious consideration of this fourth option— such as an inquiry into the benefits that might accrue from it, an estimate of its risks, or even a casual estimation of the actual losses in the way of farm income that it would entail.

In fact, numerous jurisdictions, including the Province of Ontario, have acted on concerns over excessive dependence on chemicals in farming, including pesticides, by enacting programs designed to reduce such dependence. Further evidence of the Board's bias is provided by the fact that nowhere in the *Report* is there any recognition of these concerns or of the action on them that is being taken elsewhere. One would expect, if there had been any willingness to raise the issue of lessening dependence on chemical farming, and thus to entertain its fourth option, the Board would have defined that option more carefully. The issue is not simply that of *banning* herbicides, in this case alachlor and metolachlor, but of phasing them out and of putting in place programs that compensate farmers for their losses and that enable them to pursue alternatives to farming practices made unavailable by their being phased out. The fourth option, in other words, is a catch-all for a long list of policy options that gain *prima facie* plausibility from their common feature of reducing the risks associated with use of herbicides.

The point of these remarks is not to suggest that the Board ought to have recommended this fourth option to the Minister. We hope only to have established that no one undertook to establish it as infeasible. If it is at least feasible, this strengthens our suggestion that the Board acted from a bias in favour of the status quo in agriculture. What, then, lies behind this bias?

Chemical farming is, one may say, technological farming, the application of science and engineering to growing crops (and to "animal husbandry"). It is, moreover, at least until the next generation of agricultural applications of genetic engineering supplant it, high technology, the highest currently available. A bias in favour of the status quo in agriculture is a bias in favour of technological agriculture. Our final suggestion, then, is that the Board's consideration of the fourth option—or rather, its non-consideration of it—stemmed from a presupposition concerning the place of technology in our lives. The presupposition is familiar and respectable. It goes back at least to the Enlightenment. According to this presupposition, technology is a benign tool for easing the burdens of existence, satisfying our needs, bringing comfort to our lives, reducing human misery, poverty, hunger and disease. It offers *control*. And not to exercise the control that technology, in this case agricultural technology, offers is (so it is assumed) unreasonable.

In these last few sections we have made three complementary suggestions concerning the Alachlor Review Board's deliberations. These are that underlying those deliberations were three related presuppositions: a view of rationality, of the economy and of economic values, and of technology and its place in our lives. These express values or philosophical perspectives of a normative sort which are not self-evident and which, though frequently met in our society are nevertheless highly controversial. If we are right, these presuppositions shaped the Board's deliberations and its conclusions. Remove them and little basis for those conclusions remains. Many of those who disagree with the Board's conclusions reject these presuppositions and are led to disagree with the Board because of their advocacy of different views concerning rationality, the priority of economic values, and the place of technology in our lives.

We come again, then, to a view stated earlier. It is a view that stands regardless of one's attitude toward the three presuppositions. The real issues were not the factual ones that occupied centre stage in the Alachlor Review Board Hearings. The real issues were unrecognized and not discussed. Our final suggestion is that no reasonable settlement of the alachlor debate is possible without identifying and confronting these issues.

VI

Value Frameworks in Risk Analysis

In Chapter V we analyzed the reasoning by which the Board concluded that the risk-benefit standard is the appropriate one to use in determining the *acceptability* of the risks posed by alachlor. The analysis suggested that the Board was guided by three assumptions about rationality, social order, and technology, which were not shared by those parties to the debate who disagreed with the Board's recommendation that alachlor's registration be reinstated. Given the currency of these assumptions, that Board members made them is not remarkable. What is of interest is that the Board and people generally who are engaged in the sort of enterprise that occupied the Board—the enterprise of estimating risks, balancing them against benefits, and deciding on that basis whether the risks are acceptable, that is, risk analysis—imagine that they do not make assumptions of these kinds and that the credibility of their manner of proceeding depends on their not doing so.

This view of the debate is supported by our analysis of the Board's estimate of alachlor's risks in Chapter IV. We noted there two decisive moves that the Board made in assessing alachlor's risks. First, it attributed to the government a burden of proof to establish that alachlor is unsafe. Second, it used the uncertainties in the data base on which the estimate was based in a way that favoured Monsanto. In this part we want to explore the relationships between these two moves and the three underlying assumptions discussed in the preceding chapter. The outcome of the exploration will be that the Board's manner of estimating the risks of alachlor was guided by the same underlying but unacknowledged assumptions as those that guided its consideration of the issue of standards.

It is evident that the two moves are connected: because the Board assumed that the burden of proof rested with the government, it inevitably factored uncertainties in a way that favoured Monsanto and undercut the government's case. To see this, imagine a dispute between two parties, A and B, in which A is complaining of action taken by B and it is accepted that A's complaint stands unless B can justify its action (B has "the burden of proof"). Suppose that analysis of B's argument discloses uncertainties associated with the data base and estimates on which its argument depends. That fact weakens B's justification for its action; if in the absence of justification for that action A's complaint stands, the uncertainties in B's position must work in A's favour.

In our case, A is Monsanto; B, the government. In the Board's view, the government has the burden of proof: unless the government can establish that alachlor is unsafe, it should be re-registered. Uncertainties in the risk estimation—for example, in the calculations on which estimates of applicator exposure to alachlor are based—make it more difficult for the government to rise to this challenge. Therefore, because the government is carrying the burden of proof, the uncertainties work against it and in Monsanto's favour.

Obviously, there are alternatives. (1) The Board might have decided, as the government and environmentalist groups argued it was required to do by the PCPA, that Monsanto bears the burden of proof. In that case, whatever uncertainties the risk assessment of alachlor turned up would strengthen the government's case and weaken Monsanto's. Or, (2) it might have decided that the idea of a burden of proof does not apply, that the two sides are equally bound to make their case. In these circumstances, uncertainties in the data and estimates that support either side do not strengthen the position of the other side; they just contribute to a finding of "don't know." That is, the uncertainties in the data and estimates have the effect of reducing one's confidence in whatever one concludes. At the limit, where the uncertainties are extreme, the effect is to render judgement impossible: one cannot conclude that alachlor is safe, nor can one conclude that it is unsafe. Of course, had the Board adopted this alternative, it still would have had to decide how to handle the regulatory problem. It would have had to conclude either that there was no basis for its second-guessing the Minister's cancellation of alachlor's registration (thus giving the burden of proof by default to Monsanto), or that the Minister had no basis for its regulatory action (thus giving the burden of proof by default to the government).

How are we to explain the Board's implicit placing of the burden of proof upon the government rather than Monsanto, with the corollary implication that the uncertainties in the risk estimation should work to strengthen Monsanto's case? (We say "implicit" to register the fact that the Board did not appear to be aware that it was making the moves; evidently, it made them without reflecting on what it was doing.)

How the Board's Value Framework "Framed" Its Deliberations

The role of the pro-technology and liberal assumptions in prompting these views is fairly obvious. With regard to the first, alachlor is, in effect, a technological device; the recommendation that its registration be reinstated is a proposal to exploit a technological achievement. By contrast, in its negative aspect, the government's position essentially involves denying to Canadians the benefits that such exploitation can bring.

If we assume a pro-technology bias on the Board's part, then, we can readily understand its placing the burden of proof on the government and treating the uncertainties met when estimating the risks of alachlor in a way that favoured Monsanto. A pro-technology bias translates into an assumption that a proposal to exploit a technology is *prima facie* sound, whereas resistance to such exploitation must be proved. Thus, the burden of proof was placed on the government. This argumentative strategy having been adopted, the Board looked to the evidence concerning the risks of alachlor. The uncertainties in the evidence made it difficult if not impossible for the government to offer a scientifically credible proof that the risks of alachlor were as great as it feared they were. Since the government could not rise to the challenge, the Board decided in Monsanto's favour.

The liberal assumption could only reinforce this result. "Liberalism" here means a bias against governmental interference and in favour of private individuals and groups going their own way unless their doing so would harm others. To hold this bias then is to assume that the burden of proof always rests with government, the regulator. What must be proved is that the actions of those private individuals or groups *is* likely to harm others. Failing this, interference or regulation is seen as unjustified.

But in the present case the proposal to re-register alachlor comes from the private sector, whereas the position that alachlor is unsafe and should not be registered is taken by the government. Consequently, the government bears the burden of proof and Monsanto's position is seen as sound unless scientifically convincing evidence can be advanced to refute it. Thus, the outcome is a special view of the uncertainties met when estimating the risks of alachlor. This is that the uncertainties weaken the government's case; they do so by making it difficult if not impossible for the government to rise to the challenge of defending with scientific rigour its estimation of those risks.

Finally, these references to the role of appeal to "scientific rigour" point up a way in which the Board's commitment to a distinctive view of rationality shaped its deliberations. It assumed that what cannot be established with scientific rigour is simply not credible. Coupled with the assumption that a burden of proof lay on the government, this high standard influenced the deliberations in the same way that the presence of uncertainty did, to strengthen Monsanto's position and weaken that of the government. Although the government could offer what generally would stand as good reasons for concluding that alachlor poses serious risks, on close examination its evidence failed in many respects to satisfy the standard of scientific rigour which the Board imposed. Since the government bore the burden of proof, its apparent inability to rise to the challenge contributed decisively to the Board's final recommendation.

Risk-Benefit Analysis and the Board's Value Framework

It might be thought that these remarks concerning the influence of the three assumptions merely point to idiosyncrasies of Board members. If so, they tell us little concerning risk analysis, as such, but only something about a particular and possibly aberrant risk assessment. We believe, however, that it is no accident that the Board made the three assumptions. One knows of course that, although there are exceptions, the assumptions are commonly made by risk analysts. But we want to make the stronger point, that there are conceptual links between the assumptions and the practices of risk assessment and Risk-Benefit Analysis.

This is not to say that one cannot function as a fully-accredited member of the risk assessment community without making the assumptions. The connection is more subtle than that. But it is to say that there are features of the risk assessment enterprise and of Risk-Benefit Analysis that both find support from the assumptions and motivate employing the assumptions when practising risk assessment and Risk-Benefit Analysis. We want now to support this claim.

A start is made by considering the significance of the fact that the Board chose to conceptualize the issue before it under the rubric of "risk." For the Board, the issue was, "What are the risks of using alachlor?" and "In view of the risks posed by *not* using alachlor, are the risks associated with its use acceptable?" It is important to note that this manner of construing the issue, although insisted on by Monsanto, was not forced on the Board by its mandate. Alternative ways of defining the issue were available to it. It might have adopted a legalistic approach: in reaching its decision to cancel the registration of alachlor, did Agriculture Canada follow approved procedures? Especially, did it operate within the mandate defined by the relevant legislation, the Pest Control Products Act? Or it might have conceptualized the issue as one of safety, but without translating the question,"Is alachlor safe?," into the narrower question, "Are its risks acceptable?" The latter formulation invites a decision to assume considerable risks in order to realize benefits that would be otherwise lost (or in order to avoid other risks that would otherwise be taken). To ask only whether alachlor is *safe* leaves open the alternative of employing an absolute standard, application of which might lead to pronouncing alachlor *unsafe* despite its promise of extensive economic benefits.

We note that "risk" is a Janus-faced term. One of its faces is that of the gambler. When this face is showing, to think of an activity as a risk is to invite the question whether the potential benefits from taking the risk make doing so reasonable; whether, that is, it would be a "good gamble." The other face of the concept of risk, one may say, is that of the worrier. For the worrier, in contrast to the risk-taker, to conceptualize a situation as one that presents "risks" is to find a *prima facie* reason for avoiding it.

A result of these two faces of the concept of risk is that when a situation is conceived in the terms, "Is the risk acceptable?" the individual or group confronting the question has to decide whether to approach the issue as a *risk-taker* or as one who is *risk-aversive* (or, the term we prefer, risk-cautious). The gambler face of the concept of risk suggests that this issue should be settled in favour of the risk-taker. When this face is showing, there will be a bias in favour of taking risks. This translates into the idea that a burden of proof falls on the individual or group who are risk-cautious. One who wants not to take the risk, despite a showing that doing so would be a "good gamble," needs to point to some feature of the situation that makes caution reasonable. We discuss below what some of these reasons might be.

By contrast, the aversive face of the concept of risk, which focuses on the potentiality for loss presented by a risky situation, suggests that the issue should be settled in favour of risk caution. There is a bias against taking certain kinds of risks. And this translates into the idea that a burden of proof falls on the individual or group who, despite the possibility of loss, would take the risk. In this case, it is the risk-taker who needs to justify adopting a gambling stance.

With this distinction in mind between two approaches to risk analysis, that of the risk-taker and that of the risk-cautious, the conceptual links between the three assumptions and risk analysis come into view. We ask two related questions. First, what motivates one who undertakes to estimate the risks of a situation and to determine whether the risks are acceptable to approach the task as a risk-taker? We shall argue, what by now should seem fairly obvious, that the most natural source of this decision to approach the task as a risk-taker is the frame of mind or perspective formed by the three assumptions, the individual's value framework. Second, given that one does adopt the stance of a risk-taker and is guided by the three assumptions, how does that influence the direction that the risk analysis will take? Because these two questions run into one another, the discussion that follows will address both at once.

For our purposes the importance of the discussion results from the fact that professional risk analysts typically *do* adopt the risk-taking stance. Consequently, any account of the assumptions that underlie their doing so carries implications for our understanding of the enterprise of risk analysis. In addition, typically, those lay members of the community and spokespersons for environmentalist groups who form the "opposition" in technology debates are risk-cautious. And, not incidentally, as we shall see, they also reject important elements of the frame of mind or perspective formed by the three assumptions.

Risk-Taking and the Three Assumptions

Why do we say that what motivates entering a technology debate as a risk-taker is adoption of the three assumptions, or holding the frame of mind or perspective formed by the three assumptions? For the most part, the answer is contained in our remarks above concerning the contribution of the three assumptions to the Board's "decision" that the government bears the burden of proof. Here those remarks are adapted to support the more general claim that the assumptions are conceptually linked to risk analysis when risk analysis is practised in the risk-taking mode.

First, consider again the role of a positive attitude toward technology. We have identified this as a willingness, even enthusiasm, for using technology both to ameliorate the human condition and to solve problems that arise from unanticipated side effects of deployment of technology for that purpose. Obviously, those possessed of (or, as is often the case, "by") such enthusiasm will not be detracted by evidence that going ahead carries risks. This perspective has become enshrined in the oft-repeated dictum that every action has risks associated with it. In this way, the enthusiasm referred to brings one to the porch, as it were, of the risk-benefit standard. One wants to deploy the technology, and will do so unless there are compelling reasons for restraint. The enthusiasm implies sensitivity to benefits achievable only if the technology is deployed and establishes a mind-set biased in favour of such deployment. The case for restraint needs to be made out, else, by default, the reasonable course will seem to be that of pursuing the benefit the technology offers. Finally, it is natural to expect that this case for restraint must consist in a showing, not merely that there are risks, but that these outweigh the benefits.

A liberal outlook on social and political order further specifies the sensitivity to benefits established by a pro-technology attitude. Here "benefits" refers to the standard sort of economic benefits explicitly brought into risk-benefit calculations—increased productivity, efficiency, profits and wages—but also to the benefits of freedom and non-coercion that are jeopardized by government regulation. If a technology, such as alachlor, is regulated, then benefits of all of these sorts are lost or reduced. Consequently, a liberal outlook reinforces the effect a pro-technology attitude has of motivating the risk analyst to adopt a risk-taking stance. Indeed, "motivating" is too weak a term to express the connection: were one to *profess* a liberal and pro-technology outlook but nevertheless adopt a risk-cautious rather than risk-taking stance, it might well be concluded that the profession is hollow or self-deceptive and that the individual is actually motivated by other values than those professed. That is, we are pointing here to conceptual and not merely psychological relationships.

But decisive as the pro-technology and liberal outlooks are for the risk-taking stance in risk analysis, the risk analyst's conception of rationality is even more central. We have characterized this as the conception of *instrumental* rationality. One dimension is a conviction that being rational involves being neutral, "scientifically rigorous," and precise in the way quantification and algorithms bring precision. Another, related dimension, the one we want to focus on now, is that in practical matters the rationality of a course of action is a function of its consequences—in a broad sense, utilitarianism. An action is rational if and only if it is instrumental or useful, that is, if it is productive of good consequences, "benefits."

The pro-technology and liberal attitudes contribute to one's understanding of what a "good consequence" or benefit is, and thus bring a distinctive focus to instrumental rationality. They yield preoccupation with economic benefits and technological fixes. But, more generally, concentrating on consequences institutes the perspective that the action one is evaluating promises benefits and, because there is inevitably a down-side, carries costs. The action is rational and to be followed only if the benefits outweigh the costs, or, to be more, precise, if its benefit-cost ratio exceeds those of the alternatives.

Utilitarianism of this kind is in any case a maximizing theory and when the probabilities of outcomes are included in the account the result is the principle that we are to maximize net expected utility. An important feature of this principle is its assumed comprehensiveness. It rules out as irrelevant considerations which people who understand rationality in a different way find of the highest relevance. Examples include non-consequentialist considerations such as those concerning the *justice* of distributing benefits in a particular way (for example, by recognizing the importance of the fact that those who bear the risks are not the ones who receive the benefits, and, frequently, are disadvantaged to boot), or recognition that voluntarily assumed risks may be more acceptable than involuntarily imposed ones, regardless of the offsetting benefits of the latter.

The connection of the maximizing principle with a gambling, risk-taking stance is evident. It virtually defines the stance, and in a way that is peculiarly suited to the second-mentioned dimension of instrumental rationality. If one aspires to be rational in this special sense, then one will seek always to act on the option that maximizes net expected utility, which is to say, in every case, to find the best gamble. To be risk-cautious is to reject or to qualify in some manner the maximizing principle. Thus, in the eyes of those who conceive rationality instrumentally, it is to be irrational.

Because the assumption of instrumental rationality implies a risk-taking stance, it places the burden of proof on those who are risk-cautious, and this in two senses. First, there is a burden of proof to justify *being* risk-cautious, a point made earlier. Second, there is a burden of proof on the risk-cautious party to justify its view of the unacceptability of risk. One who has a bias in favour of taking risks, a bias reinforced by the pro-technology and liberal assumptions, needs a compelling reason for not taking risks, and the compelling reason can only consist in objectively reliable data demonstrating that the risk is so serious as to be unacceptable. Where such data are not forthcoming, the promise of benefits to be gained is a decisive reason for taking the risk.

Risk-cautious persons, of course, are not convinced by the gambler's argument. They see the matter differently. For example, they may reason that the benefits are not *necessary* so there is no reason to expose oneself to the increased risk (Mrs. Van Engelen). But this distinction, between benefits that are essential and those that are optional in some sense, does not fit into the utilitarian calculus of the instrumentally rational risk-taker.

Or risk-cautious persons may hold that the timing of a utility pay-off is a highly salient feature of the situation. Even though the probability-times-utility equation may favour the taking of some risk, if there is a one-in-a-thousand-year chance of the hazard occurring, risk-cautious persons take seriously the *possibility* that the hazard may occur tomorrow, and this is something they do not want to risk. Is this "irrational"? Only on the assumption of instrumental rationality.

Or it might be salient for risk-cautious persons that the risk involved in the enjoyment of a greater benefit is part of a larger *environment of risk* which exceeds acceptable limits. Something like this seems to be involved in the environmentalists' objections to alachlor. The risk of cancer posed by alachlor use is not of special concern merely because it is *cancer* (as relevant as this may be), but also because it is not an isolated risk of cancer. Rather, alachlor use poses one more risk added to the total cancer risk in the environment; the total risk is perceived as having reached an epidemic level. Calling it "epidemic" indicates an assumption that the overall risk has reached a tolerance limit for many people, so that they are not willing to accept incremental increases in the cancer-risk environment even if these are offset by some recognizable benefit. In this situation, it becomes rational to be risk-cautious when weighing the risk of cancer against the benefits of alachlor use, *even though the simple benefit-cost ratio is favourable.*

Defenders of the underlying theory at issue here, utilitarianism, respond to these various concerns of the risk-cautious in a way that reflects their commitment to instrumental rationality. With respect to the distinction between essential and optional benefits, the response is a challenge, to translate the imputed difference in kind into differences of magnitude of preferences or utilities. With respect to the concern that an unlikely untoward event will occur tomorrow, the challenge is to defend the rationality of such concern. With respect to concern regarding the larger environment of risk, the challenge is to quantify the risk presented by the larger environment, the global risk, both as it is and as it would be with the incremental risk added in. In these ways, basing their position on an assumption that instrumental rationality is the paradigm, and finding that the considerations that motivate caution if not outright risk-aversion do not comfortably fit the paradigm, risk analysts shift the burden of proof to the cautious. The pro-technology and liberal assumptions, since they highlight benefits that the cautious would forgo, play a significant ancillary role of reinforcing the risk-taking bias.

   As indicated, where one locates the burden of proof has implications for the bearing of uncertainty on a technology debate. If the
onus of proof rests with the side that recommends caution, then serious uncertainties in the risk estimation or in the risk-benefit calculation may make it virtually impossible to justify caution, and the
palm will be awarded to the risk-takers. This is what we believe
happened in the Alachlor Review Board Hearings.
   We have based our discussion in this chapter on the pivotal role of
the issue, whether to be risk-cautious or a risk-taker. We have associated risk-taking with the three indicated assumptions, and have
argued that they form major components of the value framework
that largely determined the outcome of the alachlor debate and that,
more generally, guides the manner in which risk-taking risk analysts
approach the other technology debates of our day. This latter, general claim must stand here as an hypothesis, nothing more. Conceptually, as we have seen, it seems plausible. But to confirm it other
debates will need to be studied with the same closeness that we have
studied the debate concerning alachlor.
   Two points remain to be made. First, in the interest of completeness and symmetry, we should amplify the distinction between risk-
takers and the risk-cautious. There is, one may say, a third alternative. From the standpoint of instrumental rationality, risk-takers are
rational and the risk-cautious, because they refuse good gambles, are
not. But if we define risk-takers as those who take good gambles,
then there is a third class: those who gamble, take risks, even when
mathematically considered the prospect of gain is less than the prospect of loss. For want of a better term, we shall refer to these as "irrational risk-takers."
   The most respectable example of irrational risk-taking is met in the
literature on high consequence/low probability events. It is often
supposed that when the probability of an untoward outcome is
exceptionally small it is reasonable to ignore it, regardless of the seriousness of the hazard, were it to occur. In effect, one is prepared to
gamble simply because of the extremely low probability, and regardless of the risk (in the technical sense, probability-times-consequence). In the extreme case, if the untoward outcome is the ultimate catastrophe (the loss is infinite), then regardless of how minute
the probability, the risk (again in the technical sense) will be infinite
and it must be a bad gamble to risk it. Irrational risk-takers, however, impressed by the extreme unlikelihood of the hazard occurring,
would take the risk despite its being a bad gamble.
   The risk-cautious respond to high consequence/low probability
events in a diametrically opposed way. Where irrational risk-takers
are driven by the small probability (thinking it so small as to be,

effectively, zero) and are undeterred by the "size" of the consequence, the risk-cautious are driven by the size of the consequence, the possibility of catastrophe (thinking it so large as to be, effectively, infinite), and their caution is not allayed by the low probability.

The important point is that neither risk-cautious individuals nor irrational risk-takers base their decision on a calculation of probability-times-consequence. When the risk-taker considers the probability so small as to be negligible, the reasoning is not that this fact drives the risk down. Rather, the reasoning is that because the probability is so small (and there are different views concerning how small the probability must be) we can act as if there were *no* chance of the event occurring. This view of the probability renders the seriousness of the hazard irrelevant—an untoward outcome that is not going to occur ought not to worry one.

Similarly, when the consequence is thought to be so serious that in no circumstances ought one to risk suffering it, the low probability of its occurrence is rendered irrelevant. One does not then refuse the gamble because it is not a good one; one decides not to base one's action on a calculation at all.

There is of course a large literature on the subject. Here we shall content ourselves with remarking that, from the perspective of instrumental rationality, since both views involve preparedness to take bad gambles, both are "irrational." It is important to note the qualification built into the clause, "from the perspective of instrumental rationality." We are not claiming that they *are* irrational, but that they are such if the conception of instrumental rationality provides an adequate account of what it means to be rational. Nevertheless, it is often those who otherwise profess instrumental rationality who have convinced themselves of the reasonableness of the irrational risk-taker's approach to high consequence/low probability events. This point is illustrated above in the reference to probabilistic risk assessments of nuclear power plants.

Since the persons we are calling "irrational risk-takers" are *extreme* risk-takers, and we have identified adherence to the conception of instrumental rationality as the source of a risk-taking (as opposed to risk-cautious) stance, it is perhaps understandable that there should be readiness to depart from instrumental rationality when confronted with high consequence/low probability events. But this departure is all the more understandable when the irrational risk-takers are not only otherwise committed to instrumental rationality but also make the other two assumptions we have been discussing, a pro-technology stance and liberalism. To be prepared to take the gamble, even if, in view of the potentiality for catastrophe, it is a bad one, may make sense to those who reflect that refusing the gamble would entail both denying ourselves one of the wonders of technology and introducing restrictions on research, development and commercial enterprise which threaten the "liberal" conception of society.

We should register here our own belief that the considerations that motivate caution are often good ones. But we also believe that for many of the purposes of life, it is reasonable to ignore remote probabilities, regardless of the seriousness of the untoward event that bears that probability. (It has been remarked that there is associated with every step we take a remote possibility of disaster; it would not be reasonable to allow ourselves to be rendered immobile by awareness of such possibilities. But the implication is not that it is *always* reasonable to ignore remote possibilities. It matters what one is considering gambling. Is it one's fortune or one's life? Or the life of one's child? And it matters too what one stands to gain—*what*, not how much.) In any case, anyone who holds either of the beliefs we have confessed to holding rejects the conception of instrumental rationality, or at least rejects the contention that it provides an exhaustive account of what it means to be rational.

The second and final point concerns the relationship among the three assumptions. We have stated that they form major components of a distinctive value framework but have said little to clarify the suggestion that they form a coherent whole, much less to account for the pervasiveness of the mindset they establish. This is too large a subject to explore here but one remark will serve to round off the discussion. Each of the assumptions centrally involves the idea of *neutrality*. One may even say that each is, in a distinctive way, an assumption *of* neutrality.

Instrumental rationality is understood as value neutrality: being rational consists in finding what might be called the most efficient means for realizing ends ("values"), but reason has nothing to do with the evaluation of those ends; these, rather, are given by or along with the preferences of concerned individuals. The course of action that efficiently satisfies those preferences is rational, regardless of what those preferences specifically are.

Liberalism, from one central perspective, is the theory of the neutral state. The idea is that we do not want government to identify a preferred set of values and then mandate those for all citizens. Rather, the role of government is to protect the citizens from one another, and in some way ensure that all have access to the minimum level of goods and services that they require to pursue their goals, whatever these may be. The two kinds of values we have associated with the liberal outlook, economic values and the value of freedom (non-interference by the state) are explicable on the (questionable) assumption that neither rests on a contentious view concerning the ends that citizens ought to pursue. Thus, it is assumed, economic well-being is a universal means, needed by people regardless of the specific ends they choose to pursue, and non-interference is a condition for respecting the ends citizens adopt, whatever these happen specifically to be.

Finally, with respect to technology, the assumption of neutrality shows in the standard view that technology is a neutral and value-free tool for achieving ends: whether in a particular application it is evil or benign depends altogether on the purposes non-technologists (politicians, business people, citizens) put it to.

We are convinced that none of the three assumptions actually has the neutrality that is attributed to it. The liberal state and technology are not in fact neutral, but both specify the ends people are able to pursue in very particular ways. One need only think of the profound ways in which life in modern, technologically-based societies differs from that of the middle ages; we, of course, have more "opportunities," but these are mostly of a specific, technologically-defined sort and preclude forms of human interaction that gave medieval life its distinctive flavour.

Similarly, instrumental rationality, even when conscientiously practised, guides the researcher or decision-maker toward specific and contentious conclusions. Our investigation of the alachlor debate has illustrated this feature of instrumental rationality, and, it seems to us, the literature on liberalism and technology has made abundantly clear the impossibility of their delivering on their pretensions to being value-free and neutral.

The association of neutrality with the three assumptions helps to explain the stalemate in the technology debates of our day. It leaves one party to the debates, represented by risk analysts in a risk-taking mode, convinced that their position is supported by reasoning and premises that are value-free. The assumption of neutrality blinds them to the ways their conclusions are dependent on and presuppose a distinctive value framework. This blindness discourages, even obviates, opening up the debates to the issues that are truly divisive. These, as our analysis of the alachlor debate has shown, are non-technical issues that essentially concern values. The first condition for overcoming stalemate is to recognize the relevance and centrality of these non-technical and value-laden issues.

Conclusion

The principal lesson we derive from our study is *not* that risk analysts need to open up a global debate on the meaning of rationality and the merits of technology and the liberal state. Probably these issues are too broad for meaningful resolution. This is not to say that wider reflection on the strengths and weaknesses of the three assumptions would not help. But the assumptions show up in risk assessment and Risk-Benefit Analysis in specific ways, as assumptions about where the burden of proof lies, the appropriate response to uncertainty, the importance of the goal of protecting public safety, the weight to be assigned to economic benefits, the appropriateness of trading off safety for economic benefits, etc. In this way, the assumptions establish a distinctive way of framing risk assessment and Risk-Benefit Analysis.

These are among the more specific (value) issues that do need to be addressed by risk analysts and all others who participate in technology debates, if there is to be any hope of avoiding or overcoming deadlock. Sensitivity to the biases that are introduced by broad attitudes concerning rationality, technology and the liberal state should bring recognition by risk analysts that their activity is not, as they imagine, neutral and value-free. The more specific value issues will not go away. They will influence technology debates regardless of whether they are acknowledged. But unacknowledged assumptions cannot be confronted and conflicts concerning the risks of technologies that depend on contradictory unacknowledged assumptions cannot be resolved, at least not rationally.

Perhaps even more importantly, our identification of the way these various value frameworks inevitably feed into the risk estimation endeavour raises questions about how risk assessments and risk management should be carried out in our society. The alachlor controversy is only one example of how a technological society like ours is tempted to settle more and more of its social policy issues by turning them over to "experts" who can recommend and implement a course of action based upon an objective, even scientific, estimation. Whether it is the problem of designing organizational and management styles that maximize efficiency, designing a tax structure to maximize economic prosperity, or whatever, there are experts, with cost-benefit formulae, models of organizational behaviour, and all manner of algorithms for generating optimal policy options. The appeal of these approaches lies in part in the promise they hold out to a liberal society to be a neutral, value-free arbiter among competing communities of value in the society.

Thus, in the alachlor controversy the Alachlor Review Board held out the promise of providing a scientific solution to what appears on its face to be a political problem—whether or not Canadian society should permit a company to market a chemical and permit its farmers to put it into the soil and water. It goes without saying that to put this question into the hands of a small group of scientists is to take it out of the traditional *political* institutions where questions of social value are typically hammered out, whether these be the legislatures or the government agencies beholden to them. It is to assume that there is a rational solution to the problem which is neutral *vis-à-vis* the competing values in the community and which thus respects them all.

The significance of our analysis of the alachlor debate is its showing that the risk assessment/management process engaged in by the Alachlor Review Board does not meet the requirements of the ideal. The debate concerning alachlor's risk to its users and to the Canadian public, as well as the question of the acceptability of that risk, is *not* a purely scientific, to say nothing of a value-free or socially neutral, enterprise. It is an intrinsically *political* debate among the various value communities (e.g., industry, farmers, environmentalists, farm-wives, etc.).

As remarked above, nothing is more striking to an observer of the Alachlor Review Board Hearings than the fact that the non-scientific parties who insisted upon being heard by the Review Board were considered by Board members to be simply irrelevant to the process. The reason is quite evident. They were considered to be outside the realm of discourse which defined the Board's mandate—to review and evaluate the scientific basis for the government's cancellation decision. The public interest groups and the farm-wife, Mrs. Van Engelen, were not viewed as possessing *expertise* within the narrowly defined parameters of the Hearings. But even to the extent that these groups attempted to enter into the process on the Board's terms—by bringing forward their own data and their interpretation of data brought by others—they were treated as adversaries to the process. In our view, this was not because they did not understand the science, but rather because there was an implicit awareness among Board members that these groups were outside the *value framework* that was informing the process.

If this is true, it suggests that debates about acceptable risk in our society are intrinsically political, reflecting the lack of consensus on fundamental values like those informing the alachlor controversy. If they are political debates at their roots, then to remove them from the political institutions where those debates are typically settled, into the rarefied air of quasi-scientific expertise, is really a way of excluding alternative value frameworks leaving only those compatible, in the ways we have identified, with the "scientific" mindset. It represents the hegemony of one system of values, all in the name of value neutrality and scientific objectivity.

We do not conclude that there is no need for scientifically reliable data in the risk assessment and management debates in our society. There is such a need. The more data, and the more reliable the data, the better our judgements will be, regardless of the values we bring to the data. Our point is only that we ought not be misled into thinking that our interpretation of the data will not reflect a value framework, or, even worse, that it can substitute for such a framework. On the other hand, values, no matter how conscientious or commendable, cannot substitute for ignorance or lack of reliable data.

We can conclude, however, that the real debates about what risks are tolerable in our society—whether these be risks from chemical pesticides, from the burning of fossil fuels, the fission of atoms, or the design of baby cribs—cannot be settled simply by reference to a risk estimation algorithm that gives a "rational" solution to the problem. The alachlor case should make it clear that what is at stake in these debates is not simply who has the most accurate and sober view of the facts. What is at stake is which set of values will prevail—whose view of what it means to be rational, of what kind of society we are to have, and of the many other fundamental questions about the nature of the good life with which human beings have struggled for centuries and will continue to struggle.

Glossary

AMORTIZATION: Applied here to exposure estimates. In the context of the Alachlor Review Board Hearings, to *amortize* an exposure estimate is to convert an individual's total lifetime exposure to a lifetime average daily dose. Assume that an applicator applies alachlor one day each year for 40 years, out of a 70 year life:

$$\frac{40 \text{ years of application} \times 1 \text{ day per year}}{365 \text{ days/year} \times 70 \text{ year lifetime}} = .0015$$

The fraction expresses the percentage of the applicator's life during which he or she is assumed to be engaged in applying alachlor. To fully amortize the applicator's daily exposure is to multiply the exposure incurred on a day when the individual does apply alachlor (usually expressed in mg/kg of body weight) by .0015.

CONDITIONALLY NORMATIVE: An issue is conditionally normative in case, although it is not inherently normative, resolving it requires endorsing a normative claim. Suppose the issue to be whether alachlor exposure should be amortized. The decision that it should be amortized is descriptive in form, if *any* scientific conclusions are. (Or, the decision is only "action-guiding" in a special sense: it prescribes what one must do if one is to follow the canons of "good science.") Consequently, the issue is not inherently normative. But, in view of our lack of knowledge concerning how alachlor produces its carcinogenic effects, science cannot resolve the issue. Given the scientific uncertainty, resolving the issue requires invoking some view concerning values, a normative claim. Thus, if one judges that the health of farmers is of overriding importance, then one has a reason for concluding, as HPB did, that exposure should

153

not be amortized. It is the need to appeal to the value of health (or to some other value) that makes the issue normative. The issue is said to be "conditionally" normative because if there were no uncertainty it could be decided by appeal to the facts alone, that is, without invoking values. This marks the distinction between inherently and conditionally normative issues.

INHERENTLY NORMATIVE: An issue is inherently normative if resolving it *consists in* endorsing a normative claim. Suppose the issue is whether or not to have an abortion? Resolving it consists in deciding either that one should, or that one should not. Since both are normative claims, the issue is inherently normative. This normative feature of the issue, and of the decisions one reaches in deciding it, is "inherent" in the sense that greater knowledge concerning the facts surrounding abortion will not transform the issue into an empirical one. Nor will any appeal to those facts *suffice* to resolve the issue. To resolve the issue, one needs, in addition to appeal to the facts, one or more normative premises.

NORMATIVE: Normative claims are prescriptive: they recommend, commend, condemn, etc., in contrast to claims that are *descriptive*, which describe rather than prescribe. Normative or prescriptive claims are *action-guiding*: they assert or imply that someone should act in a certain way. Descriptive claims, except in conjunction with some normative claim, are not action-guiding. Example: the alcoholic who asserts that he ought not to be drinking is making a normative claim; he is prescribing. The disengaged observer who notes, "He has taken another drink," is merely describing; she is making a descriptive claim. Value judgements (or values in the narrow sense) are normative claims. To take an example closer to home: an assessor's belief that "It would be unfair to Monsanto to include in an exposure estimate exposure that would not have occurred had the individual been wearing adequate protective clothing" is normative and expresses a value judgement.

UNCERTAINTY: We use the term in a broad sense to refer to any information gap, vagueness, or ambiguity in an issue that prevents satisfactory settlement of the issue by appealing just to the facts of the case. We distinguish between ontological and epistemological uncertainty. If one is attempting to make a prediction, then there is ontological uncertainty in case the event one is attempting to predict is not "determined" in advance, that is, in case the accuracy of the prediction depends on some intervening event that may or may not occur. In that case, the uncertainty is "in the world." (Not everyone agrees that there is ontological uncertainty.) There is an information gap. Often, ontological uncertainty can be overcome by appealing to an event's (or an event-type's) frequency of occurrence, as actuaries do, for example. In this case, although the event one is attempting to predict is not determined, it is not uncertain.

We call uncertainty epistemological when the inability to predict is owing to an information gap, vagueness or ambiguity that is removable, but not removable in the way suggested above. One gets the missing piece of information needed to make reliable prediction possible and thus fills the information gap, resolves the ambiguity, or sheds light on the matter that had been vague. Often, unfortunately, additional information increases ambiguity by suggesting previously unthought-of possibilities.

VALUE: In its broadest sense, the term value is used so that one's values are one's *preferences*. We use the term more narrowly: values are preferences which the one who holds the preference judges to denote desirable states of affairs, actions, etc. A value judgement is *prescriptive*: in making a value judgement one identifies a state of affairs as one that *ought* to obtain, that would be prefer*able*. (The need for the distinction between the narrow and broad senses of "value" arises from the fact that sometimes what one prefers ["value" in the broad sense] one would not actually recommend ["value" in the narrow sense]. Example: an alcoholic who is trying to stop drinking may on a particular occasion crave a drink above everything else; he or she would thus "prefer" drinking to anything else. But even while preferring a drink, the alcoholic rejects the preference. The alcoholic thus values drink, in the broad sense, but disvalues it in the narrow sense.

VALUE FRAMEWORK: Generally, a person's values fall into a pattern. They form a system or framework. Some of these values are more central to the framework than others; the individual is more deeply committed to them and would be more reluctant to relinquish them. Other values are peripheral, are tacitly assigned less priority, and could be dropped more easily. Part, at least, of an individual's sense of who he or she is is provided by that person's value framework, most especially by those values that are central to that framework. Consequently, to suppose a person doesn't have a value framework would be to suppose the person didn't have a sense of who he or she is, a self-definition. In the text we have attempted to identify *part* of the value framework of certain risk assessors: liberalism, pro-technology, instrumental rationality. The latter is said to be a value because it is action-guiding: it expresses a definite view concerning how a person should behave when trying to be "reasonable."

Institutions and organizations have value frameworks as well. Often, an individual's value framework reflects that of the institution the individual is associated with. The individual's having the value framework then results from his or her having identified with that institution, or having internalized its value framework.

Notes

1. The term "mandated" science is taken from a recent book by Liori Salter, *Mandated Science* (Dordrecht: Kluwer Academic Publishers, 1988).
2. For an excellent discussion of the difference between laboratory science and mandated science, see Salter, *Mandated Science*.
3. By "classical risk assessment" we refer to the conceptualizations of it put forward by people like William Lowrance in his book *Of Acceptable Risk* (Los Altos, California: Kaufmann, 1976), and W. D. Rowe, *An Anatomy of Risk* (New York: Wiley, 1977). Also see the work of Chauncy Starr and Chris Whipple, "Risks of Risk Decisions," *Science* (1980), pp. 208, 1114-19, and, with R. Rudman, "Philosophical Basis for Risk Analysis," *Annual Review of Energy* (1976, pp. 1, 629-62); National Research Council, *Risk Assessment*, in *The Federal Government: Managing the Process* (Washington, DC: National Academy Press, 1983); and William Ruckelshaus, "Risk, Science, and Democracy," *Issues in Science and Technology* (Spring, 1985).
4. There is a terminological problem here. In the United States, the stage at which one attempts to estimate the level of risk is often called risk "assessment," In Canada, a distinction is often drawn between risk "assessment" and risk "estimation." In this Canadian usage, risk assessment includes risk estimation, *determination of the level of risk*, but includes as well certain other activities. However, there seems to be no consensus concerning what these other activities are. The issue is complicated by the fact that in Canada as well as in the United States, in certain contexts the term "risk assessment" is standardly used even when one is referring to risk estimation. For example, people who specialize in estimating risk are called risk assessors, and the field in which they work is called risk assessment. In Canada we speak of "probabilistic risk assessment" and "quantitative risk assessment" when what we have in mind are kinds of risk estimation. Given this terminological confusion, there is no clear principle to employ when deciding whether to use the term "assessment" or "estimation" at each of the points where one or the other term is called for. For the most part we shall adhere to Canadian practice by referring to "risk estimation" when we specifically have in mind the attempt to determine the level of risk. But often it will be more natural to refer to "assessment" rather than "estimation." All things considered, readers will not be misled if

157

they understand the term "assessment" in the following pages to be a synonym for "estimation." The reason is that our main thesis concerning the role of values and value frameworks focuses on their role in risk *estimation*, the process of determining the level of risk.

5. *Alachlor Review Board Hearings* (Ottawa, ON: Alachlor Review Board, 1987).

6. *Report of the Alachlor Review Board* (Toronto, ON: Atchison & Denman Court Reporting Services, 1986, 1987).

7. Here and elsewhere, as a matter of convenience, we write as if uncertainty were typically owing to "information gaps," which might suggest that eliminating the uncertainty is simply a matter of filling the gaps by doing more research. In fact, much uncertainty results not from too little, but, as it were, from too much research. Further research often results in more rather than less uncertainty.

8. For a fuller account, see the Glossary.

9. Strictly speaking, this entire argument between the Crown and Monsanto about whether the Minister was required to consider "merit and value" as well as "safety" is simply misguided. Both parties accept the standard definition of "safety" as "acceptable level of risk." The question of safety is a question of how one establishes the acceptable level of risk. In this instance the debate between the two parties is a debate about whether this determination of acceptable risk requires a reference to offsetting benefits. It does not help to say that the issue is "safety" alone. What the government seems to have *meant* here is that the issue of acceptable risk can be established by reference to *level of risk* alone.

10. A very similar value framework guided the risk assessment defended by Monsanto. HPB's value framework resembled the Review Board's in important respects, but was differently focused owing to its mandate, health protection. In the following pages we concentrate on the Review Board's value framework and devote considerably less attention to Monsanto's and HPB's.

11. Numerous authors have argued, as we do here, that risk estimation is not a value-free enterprise. Generally speaking, they fall into two groups. Some, looking to specific examples, represent the presence of values in the assessments as a sign that they are flawed. These critics, then, implicitly accept the ideal held forth by the classical view. They assume that value-free risk estimates are both possible and desirable; their contribution is to show that the ideal is seldom achieved. Others hold that the ideal is impossible of achievement but base their argument on some general philosophical position, for example, the sociology of knowledge or a critique of the "fact-value" distinction. Our approach is similar to that taken by the first group (and unlike that taken by the second group) in that we do not argue from general principles but base our conclusion on an examination of particular cases. It is similar to that taken by the second group (and unlike that taken by the first group) in that we do not regard value-free risk estimates as possible. Nor do we regard estimates which fail to be value-free as necessarily flawed. For a view concerning risk assessment which is in many respects similar to our own, see Brian Wynne, *Risk Management and Hazardous Waste* (Berlin: Springer-Verlag, 1987), and, with L. Aitken, "The System's Ability to Learn: Some Basic Problems in Post-hoc Accident Assessment," in *Policy Responses to Large Accidents*, B. Segerståhl and G. Krömer, editors (Vienna: IFASA, 1989), 185-205. But see also P. Thompson, "Risk Subjectivism and Risk Objectivism: When are Risks Real?" *Risk* 1 (1990); K. Shrader-Frechette, "Scientific Method, Anti-Founda-

tionalism, and Public Decisionmaking," *Risk* 1 (1990); Thomas McGarity, "Substantive and Procedural Discretion in Administrative Resolution of Science Policy Questions: Regulating Carcinogens in EPA and OSHA," *Georgetown Law Review* 67 (1979); Nicolas Ashford *et al.*, "A Hard Look at Federal Regulation of Formaldehyde," *Harvard Environmental Law Review* 7 (1983); Sheila Jasanoff, "Contested Boundaries in Policy-Relevant Science," *Social Studies of Science* 17 (1987); Alvin Weinberg, "Science and its Limits: The Regulator's Dilemma," *Issues in Science and Technology* (Fall 1985); Alvin Weinberg, "Science and Trans-Science," *Minerva* 10 (1972); Sheila Jasanoff, *Risk Management and Political Culture* (New York: Russel Sage, 1986); Ronald Brickman *et al.*, *Controlling Chemicals* (Ithaca, NY: Cornell University Press, 1985); and Nicholas Rescher, *Risk: A Philosophical Introduction to the Theory of Risk Evaluation and Management* (Washington, DC: University Press of America, 1983). The Weinberg and McGarity papers place particular stress on the ways uncertainty in risk assessments leads to importation of assessors' value perspectives; in this way they anticipate what we shall refer to as "conditionally normative issues" in risk estimation.

12. This is not to say, however, that narrow biases and special pleading did not play a part in the debate. But while such manoeuvres must be acknowledged, it is important not to be distracted by them, else the important message concerning risk assessment that the alachlor debate conveys will be missed. The particular biases introduced by the different parties' estimates of the risks of alachlor may be avoidable; but *no* estimate of alachlor's risks can avoid making decisive value judgements.

13. Some hold that there are no "straightforwardly scientific issues" in the present sense; that is, no issues that are entirely factual or empirical and that can therefore be settled without appeal to values or norms. We think that there are or well may be such issues, but the conclusions we reach in this study do not depend on this opinion. It is worth adding that some readers of the following pages may conclude that our position commits us to (either presupposes or implies) the view that there are no straightforwardly scientific issues. We emphatically deny that our position presupposes or implies this.

14. We recognize that for many this insistence on the presence and inevitability of inherently normative issues will seem counter-intuitive if not wrong-headed. One naturally wants to object that it is a mistake to conceptualize these issues as normative. In Chapter IV we consider this objection at length and indicate why we reject it.

15. Monsanto's estimate was guided by a value framework roughly identical to that sketched here. HPB's was more complicated, primarily owing to the fact that its institutional position dictated that it priorize the value of human health.

16. As we note elsewhere, the values risk assessors invoke often reflect the institutional context in which they operate. With that in mind, one might speak also of an *institution's* value framework. Where it is this framework that the risk assessor brings to the assessment, the personal confrontation referred to here results from the assessor's having internalized that framework.

17. For further discussion of ontological and epistemological uncertainty, see the Glossary.

18. In this case, the decisions concerning where the burden of proof should be located, like the parallel decisions concerning the kind of science to employ, seem to have been made implicitly, without recognition that they have been made. In reading the *Report* and the *Hearings*, one simply notices that assessors

approach the task by rejecting claims by one side or the other on the basis that those claims have not been fully substantiated. By proceeding in this way, the assessors effectively decide which side bears the burden of proof; whether they are aware of doing this is another matter.

19. The Board argued as follows: no human data exist to allow the extrapolation of animal to human test results. The uncertainties involved make the relationship between specific chemicals and human cancer impossible to quantify precisely. "The margins of safety used to offset these uncertainties aggregate in a multi-plicative fashion. . . . [T]hese techniques are only as reliable as the assumptions upon which they are based, and their apparent precision can result in a mask-ing of uncertainties" (*Report*, p. 29).

   Thus, because the causal data needed to calculate precisely the relationship between alachlor exposure and human cancer did not exist, the Board felt that quantitative modelling would be deceptive. It argued therefore that the judge-ment of those most experienced in inferring relationships of this kind would be more accurate—i.e., more scientific—than the multiplication of numerical estimates, and that this judgement would be more capable of including—and would in fact reflect—the important inherent uncertainty.

20. The expansion of this argument, by Dr. Farber, went as follows. Without a quantified basis for comparison, there is no rational basis for a decision con-cerning the relative toxicities of alachlor and metolachlor. In the absence of sci-entific data on humans, judgements on the relative safety of these two chemi-cals for humans must be based on "some sort of chance, whim . . . or feeling" (*Hearings*, pp. 3949, 3309).

21. The government considered the mouse feeding study to be positive with respect to the carcinogenicity of alachlor because of the increased incidence of tumours in the females compared to the female control group. Monsanto and the Board considered this study to be negative because the female control group had fewer tumours than the male control group, and the level of female tumours was no higher than that found in the male control group and the male feeding group.

22. "It may be a benign tumour, but I would not particulary like to have a benign tumour in my nose, never mind a malignant one. I think from the point of view of occupational exposure, you have to determine whether in fact you are inducing a tumour or not. We do not know . . . what happens in terms of whether the adeno-mas will go to the adenoma carcinoma" (Dr. Clegg, *Hearings*, p. 3952).

23. "This tumour [the carcinoma occurring at the 2.5 dose level] is not unique to this particular dose group. It simply occurs with a limited frequency at that dose group. If one were to look at biological plausibility, not statistical proba-bilities, the fact that this tumour occurred at other dose groups, I think, sug-gests that it is a biological response" (Dr. Ritter, *Hearings*, p. 3058).

24. "[W]ould you not agree that . . . there is considerable evidence, quite good evi-dence in some instances, that nodules can be a precursor for cancer in the liver?" When put this question by Dr. Farber during the Hearings, Dr. Clegg answered "Very definitely" (*Hearings*, p. 3939).

25. For example, when Dr. Plaa suggested that because the studies reviewed by HPB showed that since metolachlor caused liver tumours in rats, it too should be suspected to be carcinogenic, Ritter agreed (*Hearings*, p. 3170).

26. In the Hearings, Monsanto counsel Hughes told the board that because the rat is not a good metabolic model for human beings, the evidence of toxicological activity in rats should not be given great weight: "We believe that carcinogenic

or oncogenic risks to humans don't exist. . . . There is no reasonable probability that alachlor presents a risk to humans" (*Hearings*, pp. 415-20). Notice the way in which the two sides of the argument are working in different directions here. The conclusion that the risk to humans is zero does not follow from the irrelevance of the rat studies. It follows only if the rat studies are interpreted as supporting a threshold level of alachlor exposure which is carcinogenic.

27. The terms "conservative" and "liberal" are used in three different senses in these pages. In Chapter II we discussed the difference between "conservative" and "liberal" science, where conservative science is more rigorous than liberal science. We have also alluded to "liberalism" as an assumption that enters into some risk assessors' value framework. In this context, "liberalism" refers to a political point of view. And now we use the terms "liberal" and "conservative" to refer to opposing attitudes toward risk, where a liberal view is that of a risk-taker; a conservative view, that of one who is risk-aversive. (Use of these terms in the last-mentioned sense is explicable if one reflects on the connection between being "conservative" and not taking chances.) The context should make clear which of the senses of these terms we have in mind.

28. "Notwithstanding the shortcomings of estimates derived from the biomonitoring data presented by Monsanto, the Board believes that they provide a better basis than the 'patch test' studies relied upon by HPB for estimating applicator exposure to alachlor. . . . 'Patch test' studies only measure what is deposited, and must assume an absorption rate. Biomonitoring studies can measure what is excreted. When these results are coupled with the results from the dermal absorption studies reviewed by the Board, they provide more reliable estimates than 'patch tests'" (*Report*, p. 83).

29. The text of the *Report* is somewhat ambiguous, so that there is room for a different interpretation of the Board's intent than that adopted here. It seems unlikely, however, that an alternative interpretation would materially affect the conclusions we draw from our understanding of the reasoning by which the Board arrived at a worst-case exposure scenario.

30. The Board's range of both amortized and unamortized estimates was extended at the top to include estimates based on an assumption that applicators would not wear protective clothing. Evidently, the higher "bottom of the range" estimate, .0001, is used instead of the .000031 estimate because the former number, in addition to being a convenient rounding off, is roughly at the mid-point of the range of estimates for full amortization, no protective clothing.

31. Reference to amortization here illustrates the way inherently normative and conditionally normative issues blend into one another.

32. "Raw data" is not quite the term we want, since in a sense patch tests yield one set of data, biomonitoring another. By "raw data" we have in mind the amount of alachlor to which an applicator is exposed during a day in the field. The calculated exposure estimate results from first deciding whether this should be measured by a patch test or biomonitoring, then by adjusting the measured result as required by the other assumptions: absorption rate, protective clothing, amortization, amount applied and days per year, etc.

33. In referring to "bias" here we are only asserting that a particular normative stance influenced the Board's deliberations. Often, the term is used in an accusatory manner, to suggest a certain way of failing to be rational. This is not our intent. What is at issue, rather, is the *sense* in which deliberation of the sort the Board engaged in was, can, and should be rational.